AN INTRODUCTION TO THE
CALCULUS OF VARIATIONS

AN INTRODUCTION TO THE CALCULUS OF VARIATIONS

by CHARLES FOX

Late Professor of Mathematics, McGill University

DOVER PUBLICATIONS, INC.

NEW YORK

Published in Canada by General Publishing Company, Ltd., 30 Lesmill Road, Don Mills, Toronto, Ontario.

This Dover edition, first published in 1987, is an unabridged and unaltered republication of the corrected 1963 printing of the work first published by Oxford University Press, Oxford, England, in 1950. It is reprinted by special arrangement with Oxford University Press, 200 Madison Avenue, New York, N.Y. 10016.

Manufactured in the United States of America
Dover Publications, Inc., 31 East 2nd Street, Mineola, N.Y. 11501

Library of Congress Cataloging-in-Publication Data

Fox, Charles, 1897–
An introduction to the calculus of variations.

Reprint. Originally published: London : Oxford University Press, 1950 (1963 printing).
Includes bibliographical references and index. 1. Calculus of variations. I. Title.
QA315.F64 1987 515′.64 87-20112
ISBN 0-486-65499-0 (pbk.)

PREFACE

VARIATIONAL methods give us the simplest and most direct means of unifying those branches of mathematics which are commonly classified under the heading of Applied Mathematics. They are the source of such fundamental theorems as the Principle of Least Action and its various generalizations, without which a complete understanding of much of the recent revolutionary developments of Mathematical Physics is hardly possible.

Most third year students in honours mathematics should be able to master the ideas and techniques of the Calculus of Variations. During my many years of teaching at London University I felt that none of the existing texts covered the subject as I would like to teach it and so I undertook the task of writing one of my own. For the understanding of this book a knowledge of partial differentiation and differential equations will suffice.

In my opinion the value of most honours courses in mathematics would be greatly enhanced by the inclusion of at least the elements of the Calculus of Variations. But, considering the overburdened state of most curricula, this may be too much to expect at present.

After the first two chapters, which deal with the first and second variation of an integral in the simplest case, the reader can follow his own inclinations. If his interests lie in the domain of Pure Mathematics he can proceed to Chapters III and IV (generalizations, isoperimetrical problems) or to VIII, IX, and X (variable end points, strong variations). If he finds Applied Mathematics more congenial Chapters V, VI, and VII (least action, special relativity, Rayleigh–Ritz principle, elasticity) will prove more interesting.

The Calculus of Variations possesses an extensive literature, mostly of a highly specialized nature. It is perhaps unnecessary to enumerate all the works to which I am indebted because, as far as is possible, I have indicated any source from which I have borrowed by a reference in the text. In particular I should like to express my indebtedness to the classical works of G. A. Bliss, O. Bolza, C. Carathéodory, A. R. Forsyth, and J. Hadamard.

C. F.

McGILL UNIVERSITY
MONTREAL

CONTENTS

References. The equations in each section are numbered from (1) onwards. An equation in the same section as the point of reference is referred to by its number only; one in another section by its number and section number.

AN INTRODUCTION TO THE
CALCULUS OF VARIATIONS

CHAPTER I

THE FIRST VARIATION

1.1. Introduction

THE calculus of variations has ranked for nearly three centuries among the most important branches of mathematical analysis. It can be applied with great power to a wide range of problems in pure mathematics and can be used to express the fundamental principles of applied mathematics and mathematical physics in unusually simple and elegant forms.

The problem of finding points at which functions of one or more variables possess maximum or minimum values is familiar to all students of the differential calculus. In the calculus of variations we deal with the far more extensive problem of finding functional forms for which given integrals assume maximum or minimum values. In the language of geometry, we may say that this calculus deals with the problem of finding paths of integration for which integrals admit maximum or minimum values.

As a simple example consider the problem of finding the shortest distance between two points A and B, a problem whose intuitive answer is the straight line joining the two points. If s denotes the length of arc measured from A along any curve joining A and B, the problem becomes that of finding the curve for which $\int_A^B ds$ is a minimum. The calculus of variations obtains the answer by analytical methods and shows that the curves which render this integral a minimum have equations of the form $y = mx+c$. Apart from this example, intuitive answers to problems of this nature are almost non-existent. The methods of the calculus of variations therefore form a most useful addition to the domain of mathematical analysis.

Problems which can be solved by means of this calculus arose in classical times and perhaps even earlier. Grants of land, which could be completely encompassed by furrows ploughed in a specified time, were sometimes made as a reward for exceptional military or civil achievement. Thus arose the problem of finding the form which a plane curve of prescribed length must assume in order to enclose the greatest possible area (the isoperimetrical problem). Problems of this nature were dealt with largely by intuition or experiment and no progress was made towards a theoretical solution until the middle of the eighteenth century, when the researches of Bernoulli, Euler, and Lagrange first appeared.

1.2. Ordinary maximum and minimum theory

Before considering maxima and minima of integral forms we recall briefly the theory used in elementary calculus to find the maxima and minima of functions of a single variable.

Let $f(x)$ denote a continuous function of a single variable having a maximum or minimum value at $x = a$. Then for a sufficiently small ϵ we have at a maximum

$$f(a+\epsilon)-f(a) < 0, \tag{1}$$

and at a minimum

$$f(a+\epsilon)-f(a) > 0. \tag{2}$$

Taking the maximum case and assuming that $f(a+\epsilon)$ can be expanded in positive integral powers of ϵ, by Taylor's theorem, we have

$$f(a+\epsilon)-f(a) = \epsilon f'(a)+\tfrac{1}{2}\epsilon^2 f''(a)+O(\epsilon^3), \tag{3}$$

where, as usual, dashes denote differentiation. The Landau symbol O has this meaning: $O(\epsilon^3)$ possesses the property that as ϵ tends to zero the quantity $\epsilon^{-3}O(\epsilon^3)$ is bounded. From (1) and (2) at a maximum or minimum the sign of $f(a+\epsilon)-f(a)$ is independent of that of ϵ and so from (3) we must have $f'(a) = 0$. Thus the values of x which make $f(x)$ a maximum or minimum can be found by solving the equation $f'(x) = 0$.

From (1) and (3) it follows that at a maximum $f''(a)$ is negative and from (2) and (3) that at a minimum $f''(a)$ is positive. Alternatively at a maximum $f'(x)$ is a decreasing function of x and at a minimum it is an increasing function of x. Thus it is possible to discriminate quite easily between maxima and minima.

It is possible, however, that $f'(a) = 0$ and that $f(a)$ is neither a maximum nor a minimum value of $f(x)$. Such a case occurs when $f'(a) = 0$, $f''(a) = 0$, and $f'''(a) \neq 0$, and it is then customary to say that $f(a)$ is a stationary value of $f(x)$. In general all the roots of $f'(x) = 0$ are said to give rise to stationary values of $f(x)$.

These ideas, which are largely based upon common-sense notions, are fundamental in the development of the calculus of variations.

1.3. Weak variations

Our first problem in the calculus of variations will be a relatively simple one. Let

$$I = \int_a^b F(x, y, dy/dx)\, dx, \tag{1}$$

where I is a convenient symbol for the integral and F denotes a given functional form. The functional relation between y and x is not known and the problem consists in finding this relation so that I is a maximum or a minimum. In other words, given F, find the path of integration for which I is a maximum or a minimum. We confine ourselves to the case where y is a single-valued function of x in the interval (a, b).

As in § 1.2 we commence by finding the stationary values of I and then proceed to develop tests which enable us to discriminate between the cases when I is a maximum or a minimum or is neither.

Evidently the arc of integration must be of such a nature that the integral (1) can be determined; such an arc is known as an *admissible* arc. In some cases discontinuous solutions are possible, but in this book we shall confine ourselves almost entirely to continuous solutions. Subsequent analysis requires us to assume that $F(x, y, p)$ possesses partial derivatives with respect to the variables x, y, and p of at least the fourth order in an interval which includes the points $x = a$ and $x = b$. This will justify our employment of the mean-value theorem for functions of several variables. We simplify the problem appreciably by assuming that a and b, the limits of integration, are prescribed. In addition, although the functional relation between y and x is not yet known, we assume that the values of

y corresponding to $x = a$ and $x = b$, say α and β respectively, are also prescribed. Geometrically speaking the integral I must be taken along a plane curve from the given point A, co-ordinates (a, α), to the given point B, coordinates (b, β), as shown in Fig. I. 1.

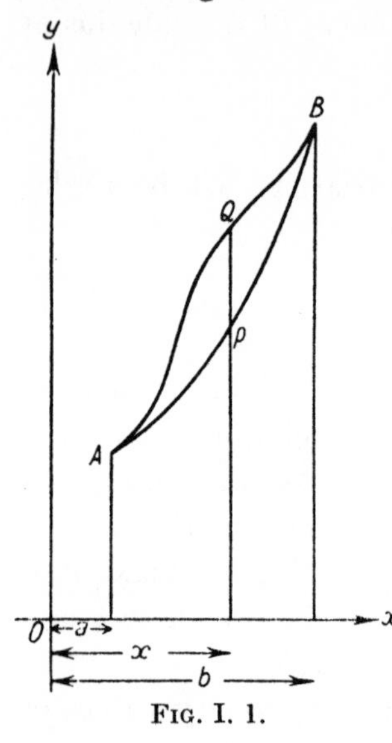

FIG. I. 1.

The problem then resolves itself into that of finding the admissible curve or curves joining A and B for which I is stationary.

Let
$$y = s(x) \tag{2}$$
be the equation of the admissible curve for which I is stationary and (see Fig. I. 1) let APB be the curve whose equation is (2). The symbol s when used to denote a functional form will always refer to the stationary case; those investigations in which s is used to denote the length of arc of a curve will be sufficiently self-explanatory to avoid the possibility of confusion. Let AQB, Fig. I. 1, be another admissible curve joining A and B and let its equation be
$$y = s(x)+\epsilon t(x), \tag{3}$$
where ϵ is an arbitrary constant independent of x and y and $t(x)$ denotes any arbitrary function of x which is independent of ϵ. With this restriction on $t(x)$ the ordinate y is said to be subjected to weak variations.† The more general case of strong variations will be dealt with later.

In Fig. I. 1 the points P and Q have the same abscissa x and $PQ = \epsilon t(x)$. Since curve (3) also passes through the points A and B we must have
$$t(a) = t(b) = 0. \tag{4}$$

† Forsyth, *Calculus of Variations*, p. 8. The variation in y is said to be weak if $t(x)$ and $t'(x)$ are of the same order of smallness. Hadamard, *Leçons sur le Calcul des Variations*, chap. ii, introduces the notion of neighbourhood to clarify the distinction between weak and strong variations.

Denoting differentiations by dashes or primes, we have from (3)

$$dy/dx = s'(x)+\epsilon t'(x). \tag{5}$$

Hence, for weak variations, as ϵ tends to zero Q tends to P and simultaneously the slope of AQB at Q tends to that of APB at P.

An example of a strong variation is given by taking

$$t(x) = \sin(x/\epsilon^2). \tag{6}$$

Differentiation leads to

$$\epsilon t'(x) = \frac{1}{\epsilon}\cos\left(\frac{x}{\epsilon^2}\right). \tag{7}$$

Evidently, as ϵ tends to zero, $\epsilon t(x)$ also tends to zero and so Q tends to P. But $\epsilon t'(x)$ oscillates infinitely and the slope of AQB at Q tends to no definite limiting direction as Q tends to P.

The distinction between weak and strong variations is of great importance, for, as will be seen later, I may admit a maximum or a minimum for weak variations but not for strong ones. Until Chapter IX we confine ourselves entirely to weak variations.

Let the value of the integral (1) when taken along the curve APB, for which it is stationary, be denoted by I_s and when taken along the neighbouring curve AQB be denoted by $I_s+\delta I_s$. Then

$$I_s = \int\limits_a^b F(x,s,s')\,dx \tag{8}$$

and

$$I_s+\delta I_s = \int\limits_a^b F(x,s+\epsilon t,s'+\epsilon t')\,dx, \tag{9}$$

where s, s', t, and t' are abbreviations for $s(x)$, $s'(x)$, $t(x)$, and $t'(x)$ respectively.

The assumption that $F(x,y,p)$ possesses continuous partial derivatives justifies an application of the mean-value theorem for functions of several variables. If the derivatives are continuous up to at least the third order we have

$$F(x,s+\epsilon t,s'+\epsilon t') = F(x,s,s')+\epsilon\left(t\frac{\partial F}{\partial s}+t'\frac{\partial F}{\partial s'}\right)+$$

$$+\frac{\epsilon^2}{2!}\left\{t^2\frac{\partial^2 F}{\partial s^2}+2tt'\frac{\partial^2 F}{\partial s\partial s'}+t'^2\frac{\partial^2 F}{\partial s'^2}\right\}+O(\epsilon^3), \tag{10}$$

where $\dfrac{\partial F}{\partial s}$ denotes $\dfrac{\partial F(x,s,s')}{\partial s}$, $\dfrac{\partial F}{\partial s'}$ denotes $\dfrac{\partial F(x,s,s')}{\partial s'}$, etc.

From (8) and (9) we finally have

$$\delta I_s = \epsilon \int\limits_a^b \left(t\frac{\partial F}{\partial s}+t'\frac{\partial F}{\partial s'}\right) dx +$$

$$+\frac{\epsilon^2}{2!}\int\limits_a^b \left(t^2\frac{\partial^2 F}{\partial s^2}+2tt'\frac{\partial^2 F}{\partial s \partial s'}+t'^2\frac{\partial^2 F}{\partial s'^2}\right) dx + O(\epsilon^3). \quad (11)$$

Denoting the coefficient of ϵ by I_1 and that of ϵ^2 by I_2, the quantities ϵI_1 and $\epsilon^2 I_2$ are sometimes referred to as the 'first variation' and 'second variation' respectively.

Evidently if I_s is a maximum then δI_s must be negative for all sufficiently small values of ϵ, whether positive or negative. Hence sufficient conditions for a maximum are $I_1 = 0$ and $I_2 < 0$. Similarly for a minimum value of I_s it is sufficient to have $I_1 = 0$ and $I_2 > 0$.

1.4. The Eulerian characteristic equation

The equation $I_1 = 0$ is easily modified to a more convenient form. Integrating by parts we have

$$\int\limits_a^b t'\frac{\partial F}{\partial s'}\,dx = \left(t\frac{\partial F}{\partial s'}\right)_{x=b} - \left(t\frac{\partial F}{\partial s'}\right)_{x=a} - \int\limits_a^b t\frac{d}{dx}\left(\frac{\partial F}{\partial s'}\right) dx. \quad (1)$$

In the term $\frac{\partial F}{\partial s'}\left(= \frac{\partial F(x,s,s')}{\partial s'}\right)$, the variables x and s are treated as constants and only the s' terms are differentiated. In the term $\frac{d}{dx}\left(\frac{\partial F}{\partial s'}\right)$, s and s' must be treated as functions of x after the partial differentiation with respect to s' and before the differentiation with respect to x.

Now it has been stipulated that $t(a) = t(b) = 0$, equation (4), § 1.3, and so the first two terms on the right-hand side of (1) vanish. The equation $I_1 = 0$ then readily reduces to

$$\int\limits_a^b t(x)\left\{\frac{\partial F}{\partial s}-\frac{d}{dx}\left(\frac{\partial F}{\partial s'}\right)\right\} dx = 0. \quad (2)$$

So far no use has been made of the arbitrariness of the function $t(x)$. We now proceed to prove that if $t(x)$ is an arbitrary function of x, then (2) can be satisfied if and only if

$$\frac{\partial F}{\partial s}-\frac{d}{dx}\left(\frac{\partial F}{\partial s'}\right) = 0, \qquad (3)$$

for all values of x between a and b.

Denote the left-hand side of (3) by $u(x)$ and suppose that $u(x)$ is not zero at all points of the curve $y = s(x)$ from A to B. Let $P_0 P_1$ (Fig. I. 2) be an arc of the curve at all of whose points $u(x)$ never vanishes, then over this arc $u(x)$ must always have the same sign, either positive or negative. Suppose it is positive, then choose $t(x)$ to be zero at all points of the arcs AP_0 and $P_1 B$ and positive at all points of the arc $P_0 P_1$. For example along $P_0 P_1$ we may take

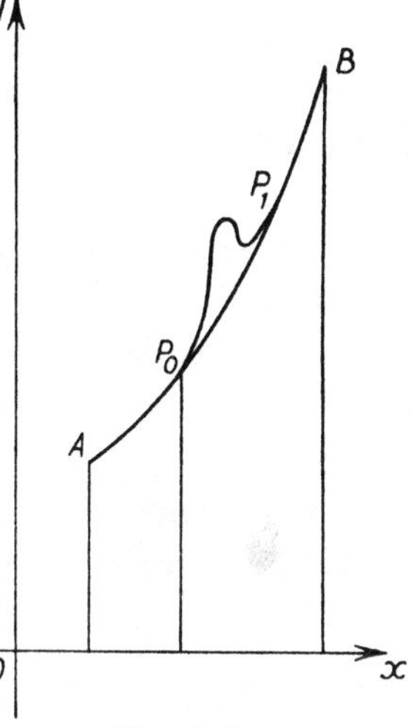

Fig. I. 2.

$$t(x) = (x-x_0)(x_1-x), \qquad (4)$$

where x_0 and x_1 are the abscissae of P_0 and P_1 respectively. But

$$\int_a^b t(x)u(x)\,dx = \int_a^{x_0} t(x)u(x)\,dx+\int_{x_0}^{x_1} t(x)u(x)\,dx+\int_{x_1}^b t(x)u(x)\,dx \qquad (5)$$

and so, since $t(x)$ vanishes in the two intervals a to x_0 and x_1 to b, we have

$$\int_a^b t(x)u(x)\,dx = \int_{x_0}^{x_1} t(x)u(x)\,dx. \qquad (6)$$

But in the interval x_0 to x_1, $t(x)$ and $u(x)$ are both positive, hence

$$\int_a^b t(x)u(x)\,dx > 0. \qquad (7)$$

But this contradicts (2), which in the terminology of this section can be written

$$\int_a^b t(x)u(x)\,dx = 0.$$

If instead of assuming that $u(x)$ is positive in the interval x_0 to x_1 we had assumed it to be negative, (7) would have been arrived at by means of similar arguments but with the inequality sign the other way. In either case the assumption that (2) does not imply (3) leads to a contradiction. Consequently, since $t(x)$ is an arbitrary function of x, the truth of (2) implies that of (3).

The discovery of (3) by Euler in 1744 inaugurated the calculus of variations in its modern form. It is a differential equation of the second order known as the characteristic equation or as Euler's equation. Its solution is the equation of a curve known as a characteristic curve or more generally as an *extremal.*

In applications of (3) we shall always replace $s(x)$ by the more convenient variable y. The results obtained may then be summed up as follows:

THEOREM 1. *The integral* $\int_a^b F(x, y, y')\,dx$, *whose end points are fixed, is stationary for weak variations if* y *satisfies the differential equation*

$$\frac{\partial F}{\partial y}-\frac{d}{dx}\left(\frac{\partial F}{\partial y'}\right)=0. \tag{8}$$

In full, this equation is:†

$$\frac{\partial F}{\partial y}-\frac{\partial^2 F}{\partial x\partial y'}-\frac{\partial^2 F}{\partial y\partial y'}\frac{dy}{dx}-\frac{\partial^2 F}{\partial y'^2}\frac{d^2y}{dx^2}=0. \tag{9}$$

The two arbitrary constants in the general solution of (9) can be determined from the fact that the extremal must pass through the two given points A and B. It is possible for the conditions to be satisfied by more than one extremal.

This theorem simplifies in two cases, case (i) when x does not occur explicitly in the function F, and case (ii) when y does not occur explicitly in the function F.

In case (i) it can easily be verified that (8) integrates to

$$F-y'\frac{\partial F}{\partial y'}=c, \tag{10}$$

† Dashes, or primes, here and subsequently, denote differentiation with respect to x, e.g.

$$y'=\frac{dy}{dx},\qquad y''=\frac{d^2y}{dx^2},\quad \text{etc.}$$

where c is an arbitrary constant. For on noting that F is now of the form $F(y,y')$ we have

$$\frac{d}{dx}F(y,y') = \frac{\partial F}{\partial y}y' + \frac{\partial F}{\partial y'}\frac{dy'}{dx}, \tag{11}$$

and

$$\frac{d}{dx}\left(y'\frac{\partial F}{\partial y'}\right) = \frac{dy'}{dx}\frac{\partial F}{\partial y'} + y'\frac{d}{dx}\left(\frac{\partial F}{\partial y'}\right). \tag{12}$$

Therefore, on differentiating (10) and using (11) and (12) we have

$$\frac{\partial F}{\partial y}y' - y'\frac{d}{dx}\left(\frac{\partial F}{\partial y'}\right) = 0. \tag{13}$$

Since y is not in general a constant, y' cannot be zero everywhere and so equations (13) and (8) are equivalent. It is thus proved that

Theorem 2. *The integral* $\int_a^b F(y, y')\,dx$, *whose end points are fixed, is stationary for weak variations if y satisfies the differential equation*

$$F - y'\frac{\partial F}{\partial y'} = c, \tag{14}$$

where c is an arbitrary constant.

In case (ii) where y is not explicit in F it is evident that the characteristic equation (8) reduces to

$$\frac{d}{dx}\left(\frac{\partial F}{\partial y'}\right) = 0, \tag{15}$$

which integrates immediately to $\partial F/\partial y' =$ constant. It is therefore proved that

Theorem 3. *The integral* $\int_a^b F(x, y')\,dx$, *whose end points are fixed, is stationary for weak variations if y satisfies the differential equation*

$$\frac{\partial F}{\partial y'} = c, \tag{16}$$

where c is an arbitrary constant.

1.5. The Legendre test

From (11), § 1.3, at a stationary value I_s of I, we have

$$\delta I_s = \frac{\epsilon^2}{2!} \int_a^b \left\{ t^2 \frac{\partial^2 F}{\partial s^2} + 2tt' \frac{\partial^2 F}{\partial s \partial s'} + t'^2 \frac{\partial^2 F}{\partial s'^2} \right\} dx + O(\epsilon^3). \tag{1}$$

Hence I_s will be a maximum (or minimum) only if the integral in (1) is negative (or positive) independently of the choice of the arbitrary function $t(x)$. If the sign of δI_s does depend upon the choice of $t(x)$, then I_s is neither a maximum nor a minimum value of I.

The detailed treatment of the integral in (1) will be postponed to Chapter II. In the remaining sections of this chapter we give a number of examples to illustrate the use of Euler's equation, (8), § 1.4, in practical cases. In order to make our treatment of these examples reasonably complete we shall anticipate a result proved in Chapter II, § 2.5. This is

THEOREM 4. *Legendre's test.*

If (i) *Euler's equation* (8), § 1.4, *is satisfied,*

(ii) *the range of integration* (a, b) *is sufficiently small,*

(iii) *the sign of* $\partial^2 F/\partial y'^2$ *is constant throughout this range,*

then I_s *is a maximum or a minimum value of* I *according as the sign of* $\partial^2 F/\partial y'^2$ *is negative or positive.*

This result is incomplete since no indication is yet given of the full range of integration permissible. A complete test can be obtained only with the help of some extensive investigations due to Jacobi which contain ideas rather too difficult for us to anticipate here. We shall therefore confine ourselves, in this chapter, to the use of the Legendre test and re-examine some of our examples in the light of the Jacobi test after it has been established in Chapter II.

The following considerations, while not forming a proof, give a certain plausibility to the Legendre test.

Since $t(a) = 0$, (4), § 1.3, we have

$$t(x) = \int_a^x t'(x)\, dx. \tag{2}$$

Hence, if M is the upper bound of $t'(x)$ in the interval (a, x), we have

$$|t(x)| \leqslant |x-a|M. \tag{3}$$

Consequently, if $|b-a|$ is sufficiently small it follows that throughout the interval $a \leqslant x \leqslant b$ the magnitude of $t(x)$ is much smaller than the upper bound of $t'(x)$. In such a case we may expect the dominant term in the integrand of (1) to be

$$t'^2 \frac{\partial^2 F}{\partial s'^2}.$$

1.6. Illustrations of the theory

EXAMPLE 1. To find the shortest distance between two given points A and B.

We shall restrict ourselves to the case where all the curves considered lie in a fixed plane through A and B and leave the more general case to a later discussion.

If APB is any curve in a fixed plane through A and B and if $s = \text{arc } AP$, then the problem resolves itself into that of finding the curves for which the integral

$$I = \int_A^B ds \tag{1}$$

is a minimum. Now

$$\frac{ds}{dx} = \left\{1+\left(\frac{dy}{dx}\right)^2\right\}^{\frac{1}{2}} \tag{2}$$

and so (1) can be replaced by the equation

$$I = \int_A^B \left\{1+\left(\frac{dy}{dx}\right)^2\right\}^{\frac{1}{2}} dx. \tag{3}$$

I is now reduced to the form given by (1), § 1.3, and Euler's equation can be applied directly to it. We have here

$$F(x, y, y') = (1+y'^2)^{\frac{1}{2}}. \tag{4}$$

Since y is not explicit in F we can use theorem 3, § 1.4, and the characteristic equation, (16), § 1.4, is then

$$\frac{y'}{(1+y'^2)^{\frac{1}{2}}} = \text{constant}. \tag{5}$$

Therefore y' is constant and by integration we obtain the family of straight lines

$$y = mx+n, \tag{6}$$

where m and n are arbitrary constants. If the coordinates of A and B are given, the values of m and n can be found uniquely.

In order to apply the Legendre test, from (4) we have

$$\frac{\partial^2 F}{\partial y'^2} = \frac{1}{(1+y'^2)^{\frac{3}{2}}} = \frac{1}{(1+m^2)^{\frac{3}{2}}}. \tag{7}$$

If the positive value of the root is taken, the sign of $\partial^2 F/\partial y'^2$ is always positive.

If the length of AB is sufficiently small, then the straight line AB is the shortest distance between A and B. But we cannot determine at present the maximum distance between A and B for which this is true. In § 2.9 the Jacobi test will tell us that the result is true for all values of the length of AB.

If instead of using theorem 3, § 1.4, use had been made of theorem 1, the characteristic equation would have been

$$-\frac{d}{dx}\left\{\frac{y'}{\sqrt{(1+y'^2)}}\right\} = 0, \tag{8}$$

which easily reduces to

$$\frac{y''}{(1+y'^2)^{\frac{3}{2}}} = 0. \tag{9}$$

This tells us that at all points on the shortest distance the curvature is zero.

Example 2. Find the shortest distance between two points A and B using polar coordinates instead of Cartesians.

With the usual notation for polar coordinates we have

$$ds^2 = dr^2+r^2\,d\theta^2 \tag{10}$$

and so the problem becomes that of finding the curve which minimizes the integral

$$I = \int\limits_A^B \left\{1+r^2\left(\frac{d\theta}{dr}\right)^2\right\}^{\frac{1}{2}} dr \tag{11}$$

$$= \int\limits_A^B \left\{\left(\frac{dr}{d\theta}\right)^2+r^2\right\}^{\frac{1}{2}} d\theta. \tag{12}$$

Either of these forms for I will do. But (11), in which r is the independent and θ the dependent variable, is more convenient since θ is not explicit and we can then use theorem 3, § 1.4. The characteristic equation

$$\frac{r^2(d\theta/dr)}{\{1+r^2(d\theta/dr)^2\}^{\frac{1}{2}}} = \text{constant} = c \tag{13}$$

is easily reduced to

$$\frac{c}{r\sqrt{(r^2-c^2)}}\frac{dr}{d\theta} = 1. \tag{14}$$

This integrates to

$$c = r\sin(\theta+\alpha), \tag{15}$$

α constant, which is the polar equation of a straight line.

An application of the Legendre test, § 1.5, shows that the conditions for a minimum are satisfied. On using (8), § 1.4, it can be shown that at all points of the shortest distance the curvature is zero, in agreement with the result of example 1.

1.7. Application to statical problems. The catenary

The calculus of variations can be usefully applied to a large variety of problems in statics.

One of the most important concepts in statics is that of potential energy. A system of bodies in a field of force possesses energy of position. This is defined as the work done by the field in moving the bodies from one configuration, P say, to a standard configuration O (O may be a configuration in which all the bodies are scattered to infinity). If the field is a conservative one, then the work function so arrived at is independent of the path from P to O and the work function is then known as the potential energy of the configuration. This is the case to which we shall confine ourselves.

W_p, the potential energy, possesses the following properties. The forces of the field can be derived from W_p by partial differentiation. If W_p is stationary then the system is in equilibrium, and if W_p is a minimum then the equilibrium is stable. For proofs of these and other properties of W_p the reader is referred to treatises on statics.†

† Routh, *Analytical Statics*, vol. i, chap. 6; Loney, *Statics*, § 175.

This stationary property of the potential energy is one to which the calculus of variations can be applied with great effect, as the following example shows.

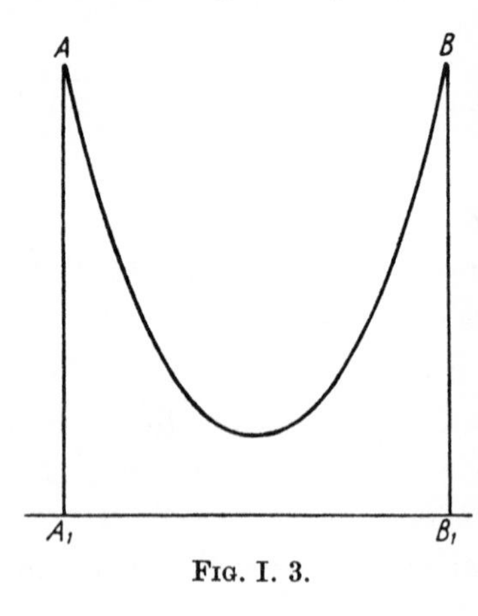

Fig. I. 3.

Example 3. Two small smooth horizontal pegs are fixed at the points A and B. The pegs are perpendicular to AB and the projections of A and B on a horizontal table are A_1 and B_1 respectively. A thin heavy uniform flexible rope, with its ends coiled at A_1 and B_1, passes over the pegs and, between A and B, hangs in equilibrium under gravity. Determine the form of the curve in which the rope hangs (see Fig. I. 3).

Evidently the potential energy of the vertical parts of the rope, AA_1 and BB_1, is constant however the curve between the pegs is varied. It can therefore be ignored in the following analysis.

If m is the mass per unit length of the rope and y is the height of the elementary arc ds above the table, then the potential energy of the part of the rope hanging between the pegs is

$$\int_A^B mgy\,ds. \tag{1}$$

Hence in a position of equilibrium this integral must be a minimum. On using $ds^2 = dx^2+dy^2$ and ignoring the constant factor mg it follows that we must minimize the integral

$$I = \int_A^B y(1+y'^2)^{\frac{1}{2}}\,dx. \tag{2}$$

Here $F(x,y,y') = y(1+y'^2)^{\frac{1}{2}}$ and x is not explicit in F, hence we may use theorem 2, § 1.4. The characteristic equation is

$$y(1+y'^2)^{\frac{1}{2}}-y'\frac{yy'}{(1+y'^2)^{\frac{1}{2}}} = c, \tag{3}$$

where c is an arbitrary constant. This simplifies to

$$y'^2 = (y^2/c^2)-1,$$

which is easily integrated to

$$y = c\cosh\left(\frac{x+b}{c}\right), \tag{4}$$

where b and c are arbitrary constants. (4) is the equation of a catenary whose directrix is the line $A_1 B_1$.

The constants b and c can be calculated if the coordinates of two points on the catenary are known, e.g. the points A and B. For simplicity let the coordinates of the pegs be $(-h, k)$ and (h, k), where h and k are positive. Then

$$k = c\cosh\left(\frac{-h+b}{c}\right) = c\cosh\left(\frac{+h+b}{c}\right), \tag{5}$$

from which it follows that $b = 0$ and

$$c\cosh\frac{h}{c} = k, \tag{6}$$

the last being an equation for c. On putting $h = cx$, equation (6) can be solved by finding the intersection of the curves

$$y = \cosh x \quad \text{and} \quad y = kx/h. \tag{7}$$

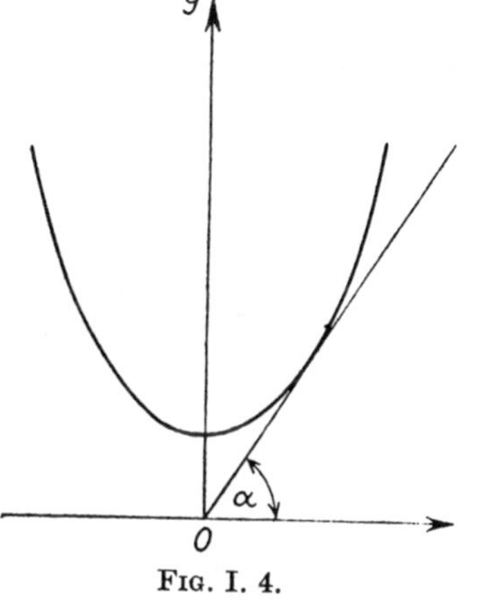

FIG. I. 4.

Fig. I. 4 shows the curve $y = \cosh x$ together with one of the tangents which passes through the origin. Denote the angle between the x-axis and this tangent by α ($\alpha = 56^\circ\ 28' = 0{\cdot}9855$ radians approximately), then the curves (7)

(i) do not intersect if $k/h < \tan\alpha$,

(ii) touch at one point if $k/h = \tan\alpha$,

(iii) intersect at two points if $k/h > \tan\alpha$.

In case (i) no catenary can be drawn through A and B having $A_1 B_1$ as its directrix, in case (ii) one such catenary can be drawn, and in case (iii) two such catenaries can be drawn.

In order to apply the Legendre test we must find the sign of

$$\frac{\partial^2 F}{\partial y'^2} = \frac{y}{(1+y'^2)^{\frac{3}{2}}}. \tag{8}$$

From (6) and (4), k, c, and y must have the same sign and so y is always positive. For the positive value of the root, it follows that I admits a minimum when the rope hangs in the form of the catenary (4), provided that the arc AB is sufficiently small. In the next chapter, § 2.13, the Jacobi test will tell us that only for the larger of the two values of c in case (iii) is there a minimum; for the smaller of the values of c in (iii) and for case (ii), (1) has stationary values which are neither maxima nor minima.

1.8. Applications to dynamical problems

The calculus of variations can be applied with great power to problems in dynamics. Consider a particle of mass m moving under gravity and let its coordinates at time t be (x, y), where x is measured horizontally and y vertically upwards. Its equations of motion are

$$m\ddot{y} = -mg, \qquad m\ddot{x} = 0, \tag{1}$$

where, as usual, dots denote differentiation with respect to t, the time. Denote the kinetic energy of the particle, $\frac{1}{2}m(\dot{x}^2+\dot{y}^2)$, by T; the potential energy, mgy, by V and let

$$L = T-V = \tfrac{1}{2}m(\dot{x}^2+\dot{y}^2)-mgy.$$

Then
$$\frac{\partial L}{\partial y}-\frac{d}{dt}\left(\frac{\partial L}{\partial \dot{y}}\right) = -mg-m\ddot{y} = 0 \tag{2}$$

by virtue of (1). But remembering that $\dot{x}$ is constant and using t as the independent and y as the dependent variable, it follows from theorem 1, § 1.4, that (2) is the characteristic equation for the integral

$$I = \int\limits_A^B L\,dt. \tag{3}$$

Thus for the motion of a particle under gravity and according to Newtonian laws of motion I is stationary.

This is a special case of Hamilton's principle, one of the most fundamental principles in applied mathematics (discovered about 1834). Hamilton's principle states that if $L = T-V$, where T is the kinetic and V is the potential energy of a system, then

$$\int_{t_0}^{t_1} L\,dt, \tag{4}$$

taken between two fixed values of the time t_0 and t_1, is stationary for a dynamical trajectory.

In more detail, let a system describe a dynamical trajectory from configuration A to configuration B under given forces (e.g. a particle P describes a parabolic path from A to B in the earth's field). Consider the case when smooth constraints, which do no work, are imposed so that the system is compelled to pass from A to B along a neighbouring path in the same interval of time (e.g. the particle P is constrained to pass from A to B by sliding along a smooth vertical curve which deviates slightly from the actual trajectory under gravity). Then (4) is stationary, and frequently a minimum, for the actual dynamical trajectory.

This principle will be proved in its most general form in Chapter V; in this chapter we shall illustrate it with a few examples. If the forces are conservative, then $T+V$ is constant during the motion and Hamilton's principle reduces to the principle of least action, which states that

$$2\int_{t_0}^{t_1} T\,dt \tag{5}$$

is stationary for a dynamical trajectory. The integral (5) is called the action. This principle was announced by Maupertuis in 1744, but the first mathematical proof was given by Euler.

Example 4. To find the trajectory of a particle moving under the earth's gravitational field.

On writing $T = \frac{1}{2}mv^2$ and $v\,dt = (ds/dt)\,dt = ds$ in (5) and ignoring the constant factor m we have to minimize the integral

$$I = \int_{A}^{B} v\,ds. \tag{6}$$

If u is the initial velocity and g denotes the acceleration of gravity, then $v^2 = u^2-2gy$, where y is measured upwards. Also $ds^2 = dx^2+dy^2$. Substituting in (6) we have

$$I = \int\limits_A^B (u^2-2gy)^{\frac{1}{2}}(1+y'^2)^{\frac{1}{2}}\,dx. \tag{6a}$$

Here $F(x,y,y') = (u^2-2gy)^{\frac{1}{2}}(1+y'^2)^{\frac{1}{2}}$ and since x is not explicit we may use theorem 2, § 1.4. The characteristic equation then becomes

$$\frac{(u^2-2gy)^{\frac{1}{2}}}{(1+y'^2)^{\frac{1}{2}}} = c \tag{7}$$

where c is an arbitrary constant. On solving for y' this is easily integrated to

$$c^2(u^2-2gy-c^2) = g^2(x-d)^2, \tag{8}$$

where c and d are arbitrary constants. (8) is evidently the equation of the well-known parabolic trajectory of elementary dynamical theory.

It is worth noting that if y is taken as the independent variable instead of x then (7) could be obtained a little more easily from theorem 3, § 1.4. This is left as an exercise to the reader.

To determine whether I is a maximum or a minimum it is best to go back to the form (5), where t is the independent variable and y the dependent one. Noting that $\dot{x}$ is constant, the Legendre test of § 1.5 then depends upon the sign of $\partial^2 T/\partial\dot{y}^2$. Now $T = \frac{1}{2}m(\dot{x}^2+\dot{y}^2)$ and so $\partial^2 T/\partial\dot{y}^2 = m$, which is positive. Hence for a trajectory with a sufficiently small arc the action is a minimum. For the maximum permissible length of this arc see §§ 2.10 and 2.14, where the Jacobi test will be applied.

If the coordinates of the end-points A and B are given, then on substitution in (8) it is possible to evaluate the two arbitrary constants c and d. In general there are two trajectories through A and B; for example if the coordinates of A are $(0, 0)$ and those of B are $(a, 0)$, the equations of the two trajectories are found to be

$$g(x^2-ax)+2yc^2 = 0, \tag{9}$$

where

$$2c^2 = u^2\pm(u^4-g^2a^2)^{\frac{1}{2}}. \tag{10}$$

This agrees with elementary theory in which it is shown that for every given horizontal range and given speed of projection there are two possible trajectories provided that $u^2 > ga$.

In its most general form Hamilton's principle contains the whole of dynamics within its scope. Not only is it elegant in theory but the ease with which the variables in the integral (4) can be changed gives it great flexibility in practice.

1.9. Applications to optical problems, paths of minimum time

Consider the problem of finding the path of a ray of light (or of a particle) which passes from A to B in minimum time. We shall confine ourselves to the case of isotropic media. If P is a point in an isotropic medium, then the physical properties of the medium at P are the same in all directions from P and are therefore functions of the coordinates (x, y, z) of P (whereas in an anisotropic medium, such as a crystal, the speed of light at P would vary with the direction of the ray). To simplify the analysis we shall also confine ourselves to plane paths, although this is not essential.

Let μ be the *inverse of the speed* and ds the element of arc at the point P, then the time taken to traverse ds is $\mu\, ds$, where μ is a function of the coordinates of P only. The problem then resolves itself into that of finding the path for which

$$\int_A^B \mu\, ds \qquad (1)$$

is a minimum. On using $ds^2 = dx^2+dy^2$ we must then minimize the integral

$$I = \int_A^B \mu(1+y_1^2)^{\frac{1}{2}}\, dx, \qquad (2)$$

where $y_1 = dy/dx$. From theorem 1, § 1.4, the characteristic equation is

$$(1+y_1^2)^{\frac{1}{2}}\frac{\partial\mu}{\partial y}-\frac{d}{dx}\left\{\frac{\mu y_1}{(1+y_1^2)^{\frac{1}{2}}}\right\} = 0. \qquad (3)$$

This equation is capable of a simple but important physical interpretation. If the tangent to the minimum path makes an angle ψ with the x-axis, then from elementary theory we have $(1+y_1^2)^{\frac{1}{2}} = ds/dx$ and $y_1/(1+y_1^2)^{\frac{1}{2}} = \sin\psi$. Equation (3) is then easily reduced to

$$\frac{\partial\mu}{\partial y} = \frac{d}{ds}(\mu\sin\psi). \qquad (4)$$

Now this equation represents a geometrical property of the minimum path which is independent of the choice of axes. Therefore its interpretation can be obtained by using any convenient system of axes. The most convenient system is chosen as follows.

Since μ is a function of x and y only, the equation

$$\mu = \text{constant} \tag{5}$$

is that of a plane curve and by varying the constant we get a family of curves which will be called *level curves*. If μ is one-valued, then through any point there will be only one level curve (if μ is many valued the same will be true if one branch of μ is adhered to and branch points are avoided). If O is a point on the minimum path, let us take it as the origin and the normal and tangent to the level curve through it as the x and y axes respectively. In Fig. I. 5 the level curve is drawn as a continuous line and the minimum path as a discontinuous line.

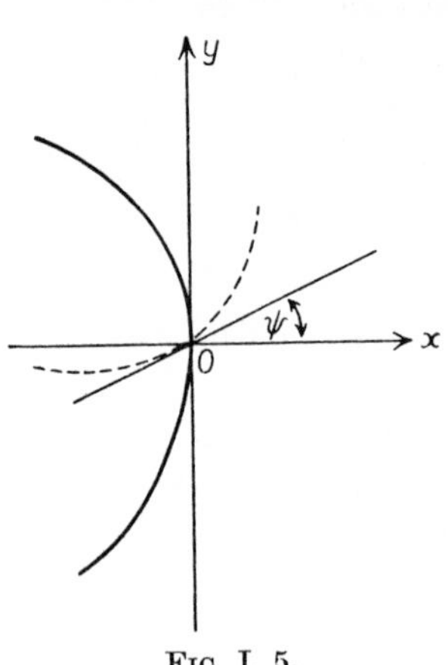

FIG. I. 5.

Now at O the tangent to the curve (5) is perpendicular to the x-axis, hence $-\dfrac{\partial\mu}{\partial x}\Big/\dfrac{\partial\mu}{\partial y}$ must be infinite and so $\dfrac{\partial\mu}{\partial y} = 0$. Equation (5) then becomes

$$\frac{d}{ds}(\mu \sin\psi) = 0, \tag{6}$$

i.e. there is no variation in $\mu \sin\psi$ on travelling a small distance ds along the minimum curve. In other words, if at a point of intersection of the level curves $\mu =$ constant and the path of minimum time the angle between the normal to the former and the tangent to the latter is ψ, then

$$\mu \sin\psi = \text{constant} \tag{7}$$

for all points along the minimum path.

For a ray of light μ, which is inversely proportional to the speed, must be directly proportional to the refractive index of the medium through which the ray is passing. Equation (7) is then a statement of Snell's law of refraction.

Conversely from Snell's law of refraction it follows that in its passage through isotropic media a ray of light moves from one point to another in stationary time.

With more elaborate analysis these results can be extended to three-dimensional space.† The Jacobi test of the next chapter will show that in general the time along an optical path is a minimum.

The Eulerian characteristic equation has so far been integrated in two cases, namely theorems 2 and 3, § 1.4. Equation (7) above gives us another case of integration. This important result is not restricted to optical theory, as the following dynamical example shows.

For a particle moving in a conservative field of force the principle of least action, § 1.9, states that $2\int_A^B T\,dt$ is stationary for the actual path. Here T (the kinetic energy) $= \frac{1}{2}mv^2$ and on writing $v\,dt = ds$, and ignoring the constant factor m the integral to be minimized becomes

$$\int_A^B v\,ds. \tag{8}$$

If V is the potential energy and h the total energy of the particle we have

$$\tfrac{1}{2}mv^2+V = h. \tag{9}$$

Thus (8) can be rewritten in the form

$$\int_A^B (h-V)^{\frac{1}{2}}\,ds, \tag{10}$$

where V is a function of position only. The results of the first part of this section then show that the curves which minimize (10) must have the following property. Writing $\mu = (h-V)^{\frac{1}{2}}$ in order to correlate integrals (1) and (10), we see that the level curves are given by $(h-V)^{\frac{1}{2}} =$ constant. This is equivalent to $V =$ constant, so that the level curves are the same as the

† Forsyth, *Calculus of Variations*, p. 258.

equipotential curves of dynamical theory. The interpretation of (7) for this case can be stated as follows:

Let the path of the particle cut an equipotential curve at the point P and let the tangents to the two curves at P intersect at an angle $\frac{1}{2}\pi-\psi$. Then

$$(h-V)^{\frac{1}{2}}\sin\psi = \text{constant}. \tag{11}$$

This result, which is true at all points of the trajectory, has been obtained without integration.

In the case of a particle moving in the earth's gravitational field $V = mgy$, and it can be shown that (11) is equivalent to the statement that the subnormal at all points of the trajectory is constant, a well-known property of the parabola.

This idea was applied to dynamics with great effect by Hamilton and enabled him to develop the theory of contact transformations.† It is also found useful in the modern theory of wave mechanics.

1.10. Geodesics on a sphere

Consider the family of curves lying wholly on a given surface, S say, and passing through two given points, A and B, both lying on S. Among these curves there will be one for which the length of the arc AB is a minimum and others for which this length is stationary. Such curves are known as geodesics. They are most easily determined by the methods of the calculus of variations, as the following example where S is a sphere, will show.

Let (x, y, z) be the coordinates of a point P on a sphere whose centre is at the origin and whose radius is a. Then in polar coordinates $x = a\sin\theta\cos\phi$, $y = a\sin\theta\sin\phi$ and $z = a\cos\theta$, where θ is the colatitude and ϕ the longitude or azimuth (Fig. I. 6). Evidently

$$ds^2 = dx^2+dy^2+dz^2 = a^2(d\theta^2+\sin^2\theta\, d\phi^2), \tag{1}$$

and so we are required to minimize the integral

$$I = a\int\limits_A^B (1+\phi_1^2\sin^2\theta)^{\frac{1}{2}}\, d\theta, \tag{2}$$

† Whittaker, *Analytical Dynamics*, chap. xi.

where $\phi_1 = d\phi/d\theta$ and the positive value of the root is taken. From theorem 3, § 1.4, it follows that the characteristic equation is

$$\frac{\partial}{\partial\phi_1}(1+\phi_1^2\sin^2\theta)^{\frac{1}{2}} = \text{constant}. \tag{3}$$

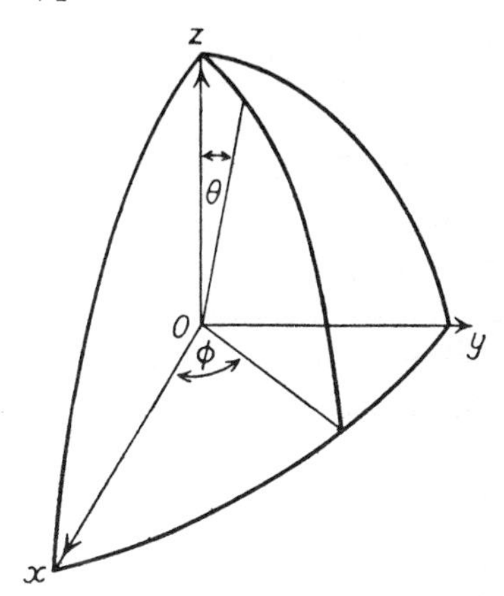

Fig. I. 6.

This differentiates to

$$\frac{\phi_1 \sin^2\theta}{(1+\phi_1^2\sin^2\theta)^{\frac{1}{2}}} = \sin\alpha, \tag{4}$$

where α is constant. Solving for ϕ_1 and integrating we get

$$\phi+\beta = \int \frac{\sin\alpha\, d\theta}{\sin\theta(\sin^2\theta-\sin^2\alpha)^{\frac{1}{2}}}. \tag{5}$$

On substituting $\theta = \tan^{-1}(1/u)$ in the integral on the right-hand side it reduces to

$$-\int \frac{\tan\alpha\, du}{(1-u^2\tan^2\alpha)^{\frac{1}{2}}}. \tag{6}$$

Equation (5) now integrates to

$$\cos(\phi+\beta) = \frac{\tan\alpha}{\tan\theta}, \tag{7}$$

which, on transformation into Cartesian coordinates, gives us

$$x\cos\beta - y\sin\beta = z\tan\alpha. \tag{8}$$

This is the equation of a plane through the centre of the sphere.

Thus the geodesics on a sphere are obtained as the intersection of the sphere and a plane through its centre, and so must be arcs of great circles.

The Legendre test of § 1.5 depends upon the sign of

$$\frac{\partial^2}{\partial\phi_1^2}(1+\phi_1^2\sin^2\theta)^{\frac{1}{2}} = \frac{\sin^2\theta}{(1+\phi_1^2\sin^2\theta)^{\frac{3}{2}}}. \tag{9}$$

Since the positive value of the root is taken it follows that I, equation (2), admits a minimum for a sufficiently small length of arc AB. The maximum permissible length of arc is obtained in § 2.9 and § 2.15, where the Jacobi test is used.

1.11. Brachistochrone

The following problem, first solved by Bernoulli in 1696, led to the foundation of the calculus of variations in its modern form.

EXAMPLE 6. A particle slides under gravity from rest along a smooth vertical curve joining two points A and B. To find the curve in which the time from A to B is a minimum. A curve of minimum time in dynamics is known as a *brachistochrone.*

Taking the upper point A as the origin and measuring y vertically downwards, the velocity at a depth y is $(2gy)^{\frac{1}{2}}$ and the time from A to B is

$$\int \frac{ds}{v} = \frac{1}{(2g)^{\frac{1}{2}}}\int \frac{1}{y^{\frac{1}{2}}}(1+y_1^2)^{\frac{1}{2}}\,dx, \tag{1}$$

where $y_1 = dy/dx$.

From theorem 2, § 1.4, this is stationary when

$$\frac{1}{y^{\frac{1}{2}}}(1+y_1^2)^{\frac{1}{2}} - \frac{y_1^2}{y^{\frac{1}{2}}(1+y_1^2)^{\frac{1}{2}}} = \text{constant}, \tag{2}$$

which simplifies to $\qquad y(1+y_1^2) = 2c. \qquad (3)$

On putting $y_1 = \tan\psi$ in (3) we get

$$y = c(1+\cos 2\psi), \tag{4}$$

and on writing $dx = dy\cot\psi$, substituting for dy from (4) and integrating we get

$$x = a - c(2\psi+\sin 2\psi), \tag{5}$$

where a and c are arbitrary constants. Equations (4) and (5) are evidently the equations of a cycloid in the usual parametric form, with 2ψ as the angular parameter and c as the radius of the generating circle. The generating circle rolls on the horizontal line through A which lies in the vertical plane through AB.

On substituting the coordinates of A, $(0, 0)$ in (4) and (5), we get $\psi = \frac{1}{2}\pi$ and $a = c\pi$. On substituting the coordinates of B in (4) and (5) and eliminating ψ the values of a and c are then easily found.

The Legendre test of § 1.5 depends upon the sign of

$$\frac{\partial^2}{\partial y_1^2}\left\{\frac{(1+y_1^2)^{\frac{1}{2}}}{y^{\frac{1}{2}}}\right\} = \frac{1}{y^{\frac{1}{2}}(1+y_1^2)^{\frac{3}{2}}}. \tag{6}$$

On taking the positive value of each root it follows that the integrals of (1) admit a minimum for a sufficiently small arc of the cycloid.

1.12. Minimal surfaces

Example 7. Given two points A and B and a line l which intersects AB produced. Let ω denote the plane through A, B, and l. To find the curve joining A and B which lies in the plane ω and which, on rotation about l through four right angles, generates a surface of minimum area.

If ds is the element of arc of a curve joining A to B and y is its distance from l, then the area generated by rotation about l is

$$2\pi \int y\,ds. \tag{1}$$

The problem of minimizing this integral has already been dealt with in § 1.7 and the solution is therefore given by (4), § 1.7. Thus (1) is minimized by the curve whose equation is

$$y = c\cosh\left(\frac{x+b}{c}\right), \tag{2}$$

where b and c are arbitrary constants. This is a catenary whose directrix coincides with the axis of rotation.

The analysis of § 1.7 applies completely to the problem of this paragraph. If for simplicity the two points A and B have coordinates (h, k) and $(-h, k)$, where h and k are positive, and if $k > h\tan\alpha$ (where $\alpha = 56^\circ\,28'$), then there are two real values of c satisfying the equation

$$c\cosh\frac{h}{c} = k, \tag{3}$$

and for the larger of these values of c there will be a minimum.

The surface generated by the revolution of a catenary about its directrix is a special case of a class of surfaces known as minimal surfaces. A more general discussion of these surfaces will be given in § 3.11.

1.13. Principle of least action. Inverse square law

EXAMPLE 8. A particle of mass m is attracted towards a fixed point O by a force of magnitude $m\mu/r^2$, where μ is constant and r is its distance from O. Show that the orbit of the particle is an ellipse, one of whose foci is at O.

On using the principle of least action, § 1.8, we must minimize the integral

$$2\int T\,dt, \tag{1}$$

where T (the kinetic energy) $= \frac{1}{2}mv^2$. On writing $v\,dt = ds$ in (1) the integral is transformed to

$$m\int v\,ds. \tag{2}$$

We first find v in terms of r by using the potential energy. This is the work done by the field in displacing a unit particle to some convenient standard position. At a point distant r from O the potential energy is

$$-\mu\int\limits_r^\infty \frac{dr}{r^2} = -\frac{\mu}{r}. \tag{3}$$

Hence, from the conservation of energy, we have

$$\tfrac{1}{2}mv^2 - m\frac{\mu}{r} = \text{constant}, \tag{4}$$

or, in more convenient form,

$$v^2 = \mu\left(\frac{2}{r}-\frac{1}{a}\right), \tag{5}$$

where a is constant. On using polar coordinates, for which $ds^2 = dr^2+r^2\,d\theta^2$, and ignoring the constant factor $m\sqrt{\mu}$, (2) becomes

$$\int\left(\frac{2}{r}-\frac{1}{a}\right)^{\frac{1}{2}}\left\{1+r^2\left(\frac{d\theta}{dr}\right)^2\right\}^{\frac{1}{2}}dr. \tag{6}$$

Taking r as the independent and θ as the dependent variable this integral can be minimized by using theorem 3, § 1.4, the characteristic equation being

$$\left(\frac{2}{r}-\frac{1}{a}\right)^{\frac{1}{2}}\frac{r^2(d\theta/dr)}{\{1+r^2(d\theta/dr)^2\}^{\frac{1}{2}}} = \text{constant} = c^{\frac{1}{2}}. \tag{7}$$

Solving for $d\theta/dr$ we have

$$\frac{d\theta}{dr} = \frac{(ac)^{\frac{1}{2}}}{r(2ar-r^2-ac)^{\frac{1}{2}}}. \tag{8}$$

On transforming to a new variable u where $r = c/u$, this is easily integrated to

$$\frac{c}{r} = 1+\left(1-\frac{c}{a}\right)^{\frac{1}{2}}\cos(\theta+\beta), \tag{9}$$

where β is another arbitrary constant. For real results we must have $c < a$.

The orbit is therefore a conic having O as a focus with eccentricity $(1-c/a)^{\frac{1}{2}}$. In (7) $c^{\frac{1}{2}}$ must be real and so c must be positive. Hence (9) is an ellipse, parabola, or hyperbola according as a is positive, infinite, or negative. On applying this result to (5) we obtain the velocity formula appropriate to each type of orbit.

The Legendre test tells us that the action is a minimum for a sufficiently small arc of the orbit. The details of the analysis are left to the reader. For the permissible length of this arc see § 2.16, where the Jacobi test is applied.

1.14. Principle of least action. Direct distance law

EXAMPLE 9. A particle of mass m is attracted towards a fixed point O by a force $m\mu r$, where μ is constant and r is its distance from O. Show that its orbit is an ellipse whose centre is at O.

By using the potential energy as in the last section, it can be shown that

$$v^2 = \mu(a^2-r^2). \tag{1}$$

On using polar coordinates, the principle of least action shows that the integral to be minimized is

$$\int (a^2-r^2)^{\frac{1}{2}}\left\{1+r^2\left(\frac{d\theta}{dr}\right)^2\right\}^{\frac{1}{2}}dr. \tag{2}$$

From theorem 3, § 1.4, the characteristic equation is

$$(a^2-r^2)^{\frac{1}{2}}\frac{r^2(d\theta/dr)}{\{1+r^2(d\theta/dr)^2\}^{\frac{1}{2}}} = c^2, \tag{3}$$

where c is an arbitrary constant. On writing $r^2 = c^2/u$ and solving for $d\theta/du$ this equation can be integrated. The details are left to the reader, the result is

$$\frac{c^2}{r^2} = \left(\frac{a^4}{4c^4}-1\right)^{\frac{1}{2}}\cos 2(\theta+\beta)+\frac{a^2}{2c^2}, \tag{4}$$

where β is another arbitrary constant. This is the polar equation of an ellipse whose centre is at O, the origin.

The Legendre test shows that the action is a minimum for a sufficiently small arc of the orbit. For the permissible length of this arc see § 2.17.

1.15. A Problem in fluid mòtion

Example 10. A uniform perfect liquid rotates inside a cylindrical container with constant angular velocity ω about a vertical axis. Show that the free surface is a paraboloid of revolution.

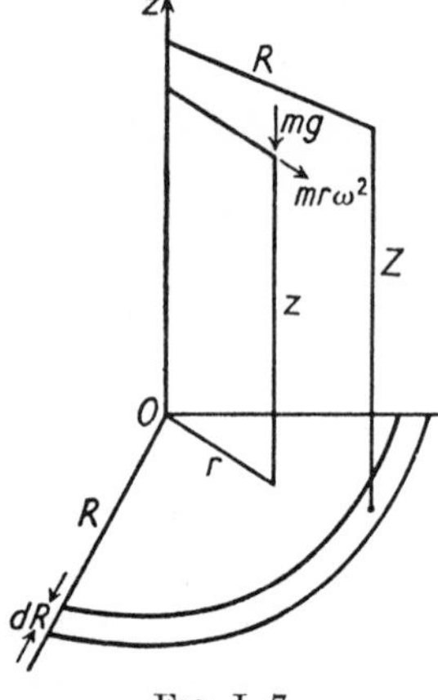

Fig. I. 7.

Let r and z be the distances of any point P of the liquid from the axis of rotation and the bottom of the container respectively. Consider the liquid as made up of a number of elementary particles of which one, of mass m, is situated at P and apply D'Alembert's principle, which states that the external forces and the reversed effective forces are in equilibrium.

For the particle at P the external force is mg vertically downwards. The reversed effective force (i) is of magnitude $mr\omega^2$, (ii) lies in the horizontal line which passes through P and intersects the axis of revolution, (iii) is directed away from the axis (Fig. I. 7). The potential function for such a system of forces is $mgz-\frac{1}{2}mr^2\omega+a$, where a is constant. We now sum this function for all particles of the liquid situated on a thin cylinder whose axis is the axis of revolution and whose internal and external radii are x and $x+dx$. The sum is $(\frac{1}{2}\rho gy^2-\frac{1}{2}\rho x^2y\omega+ay)2\pi x\,dx$, where y is the height from the bottom of the cylinder to the free surface and ρ is the density of the liquid. On summing for all such cylinders the potential energy of the system is then equal to

$$\int\limits_0^b (\tfrac{1}{2}\rho gy^2-\tfrac{1}{2}\rho x^2y\omega+ay)2\pi x\,dx, \tag{1}$$

where b is the radius of the container.

When the potential energy is stationary the system is in equilibrium and when it is a minimum the equilibrium is stable. To minimize (1) we have the Eulerian equation

$$\rho gy-\tfrac{1}{2}\rho x^2\omega+a = 0. \tag{2}$$

This is the equation of the free surface and is evidently a paraboloid of revolution.

Since there is no term containing y' in the integrand the Legendre test cannot be applied. On introducing the variation† $y = s(x)+\epsilon t(x)$, as in § 1.3, it is easily found that the second variation is

$$\tfrac{1}{2}\epsilon^2 \int\limits_0^b t^2(x)\rho g\,dx.$$

This is positive, independently of the choice of ϵ and $t(x)$, and the potential energy is therefore a minimum when the free surface is the paraboloid (2).

1.16. Newton's solid of minimum resistance

Example 11. The following problem, propounded by Newton in the *Principia* (1687), was one of the earliest to be solved by the methods of the calculus of variations.

A solid of revolution moves, with constant velocity, in the direction of its axis in a perfect incompressible liquid. If the resistance at any point is proportional to the square of the normal component of the velocity, find the shape of the surface in order that the total resistance should be a minimum.

Taking the y-axis as the axis of revolution, the positive

† The variations must be such that $t(0) = 0$ and $t(b) = 0$. A complicated mechanism would be required in order to obtain these results in practice.

direction being the direction of motion (Fig. I. 8), let v denote the velocity. If ψ is the angle between the x-axis and the tangent to the meridian curve at the point P, the resistance of the solid is proportional to

$$\int \{v^2\cos^2(\pi-\psi)2\pi x\,ds\}\cos\psi = \int v^2\cos^2\psi\,2\pi x\,dx, \tag{1}$$

the integration being taken along an arc of the meridian curve.

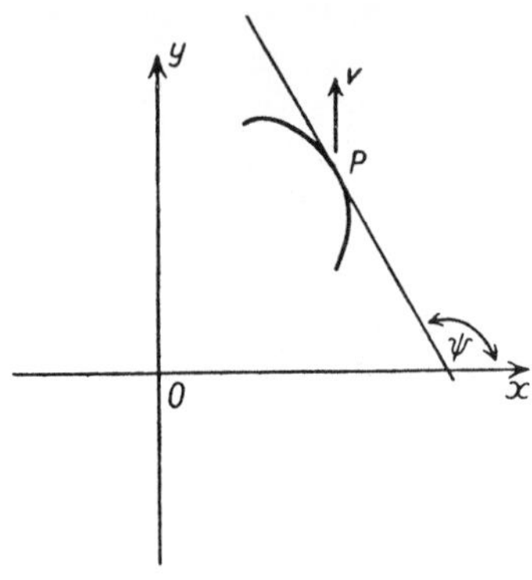

FIG. I. 8.

Since $\cos\psi = dx/ds$ and $\cos^2\psi = 1/\sec^2\psi = 1/(1+y'^2)$ where $y' = dy/dx$, it follows that we must minimize the integral

$$I = \int \frac{x}{1+y'^2}\,dx. \tag{2}$$

From theorem 3, § 1.4, the characteristic equation is

$$\frac{xy'}{(1+y'^2)^2} = \text{constant}. \tag{3}$$

This is best solved in parametric form. On writing c for the constant and $y' = p$ we have

$$x = \frac{c}{p}(1+p^2)^2, \tag{4}$$

and

$$y = \int p\,dx. \tag{5}$$

Substituting for dx from (4) and integrating (5) we obtain

$$y = a+c(-\log p+p^2+\tfrac{3}{4}p^4). \tag{6}$$

Equations (4) and (6) are the required equations, in parametric form, of the meridian curve of the solid of least resistance.

The Legendre test depends upon the sign of $\partial^2 F/\partial y'^2$ which, for this problem, is equal to

$$-2x\frac{(1-3y'^2)}{(1+y'^2)^3}.$$

Hence for a sufficiently small arc there will be a minimum if, when x is positive, $y' > 1/\surd 3$, or $y' < -1/\surd 3$. The results, however, do not agree well with experiment.

1.17. Discontinuous solutions

In the examples hitherto discussed y and its derivatives have all been continuous functions of x. Discontinuous solutions are also possible in some cases. In these y is defined for every relevant value of x and must be continuous, but dy/dx, or some higher derivative, may be a discontinuous function of x. Suppose that dy/dx is discontinuous at $x = x_1$, then, as x tends to x_1, dy/dx tends to the limit m or n ($\neq m$) according as $x-x_1$ tends to zero through negative or positive values. At $x = x_1$ the path of integration is said to have a corner. Except for the brief remarks of this section, such solutions will not be considered in this book.

Discontinuous solutions are generally obtained by joining together several continuous arcs, each of which satisfies the Eulerian characteristic equation. The corners occur at the points of junction.

We now prove that $\partial F/\partial y'$ and $F-y'(\partial F/\partial y')$ must be continuous functions of x at all corner points as well as at ordinary points.†

Differentiate the two equations

$$\frac{\partial F}{\partial y'} = \int^x \frac{\partial F}{\partial y}\,dx \tag{1}$$

and

$$F-y'\frac{\partial F}{\partial y'} = \int^x \frac{\partial F}{\partial x}\,dx \tag{2}$$

† Sometimes known as the Weierstrass–Erdmann corner conditions. The integral to be minimized is, as usual, $\int F(x, y, y')\,dx$ and we write

$$F(x, y, y') = F.$$

totally with respect to x. In each case we deduce that y satisfies the characteristic equation. Conversely if y satisfies the characteristic equation, then (1) and (2) can be deduced by integration with respect to x. From the usual continuity properties of integrals it then follows that the expressions on the left-hand sides must be continuous functions of x.

The following example illustrates how these results can be applied in practice.

EXAMPLE 12. Minimize the integral

$$I = \int (y'^2-1)^2\,dx. \tag{3}$$

From theorem 3, § 1.4, the extremals are the straight lines

$$y = ax+b. \tag{4}$$

Suppose that a corner exists at $x = x_1$ with slopes m and n $(\neq m)$ on either side. From the continuity of $\partial F/\partial y'$ we have

$$4m(m^2-1) = 4n(n^2-1) \tag{5}$$

and from that of $F-y'(\partial F/\partial y')$ we have

$$(m^2-1)^2-4m^2(m^2-1) = (n^2-1)^2-4n^2(n^2-1). \tag{6}$$

Since $m \neq n$ we may divide each of these equations by $(m-n)$, when it follows easily that $m = 1$ and $n = -1$ are solutions. Hence the minimizing path of integration consists of a series of straight lines making angles $\pm\frac{1}{4}\pi$ with the x-axis. Evidently $I \geqslant 0$ so that the minimizing path, for which $y' = \pm 1$, gives an absolute minimum.

EXAMPLE 13. Minimize the integral

$$I = \int (y'-\tan\alpha)^2(y'-\tan\beta)^2\,dx,$$

where α and β $(\neq \alpha)$ are constants.

At a corner with slopes m and n on either side, $\partial F/\partial y'$ can be continuous only if it has the same value for $y' = m$ as for $y' = n$. By Rolle's theorem† it follows that $\partial^2 F/\partial y'^2$ must vanish for some intermediate value of y'. Therefore no discontinuous solutions can exist for a region in which $\partial^2 F/\partial y'^2$ has no zeros.

† R. Courant, *Differential and Integral Calculus*, Blackie & Son, vol. 1, p. 104.

For example, consider the catenary problem of § 1.7. Here

$$\frac{\partial^2 F}{\partial y'^2} = \frac{y}{(1+y'^2)^{\frac{3}{2}}}. \tag{7}$$

Since this has no zero for any finite value of y there can be no discontinuous solutions for the problem of § 1.7.

Similar remarks apply to the least action and brachistochrone problems discussed in §§ 1.8, 1.11, 1.13, and 1.14.

1.18. Characteristic equation an identity

We conclude the chapter by discussing the case when the characteristic equation reduces to a zero identity.

Consider the problem of minimizing the integral

$$\int \{2xy+(x^2+3y^2)y'\}\,dx. \tag{1}$$

From theorem 1, § 1.4, the characteristic equation

$$\frac{\partial F}{\partial y}-\frac{d}{dx}\left(\frac{\partial F}{\partial y'}\right) = 0 \tag{2}$$

becomes
$$(2x+6yy')-\frac{d}{dx}(x^2+3y^2) = 0. \tag{3}$$

On performing the differentiation the left-hand side vanishes identically. Evidently the calculus of variations has no application to cases of this sort. Instead we must make use of the following theorem:

THEOREM 5. *If the characteristic equation* (2) *vanishes identically, then the indefinite integral* $\int F(x,\,y,\,y')\,dx$ *can be evaluated as a function of* x *and* y. *The integral has then a value which is independent of the path of integration and which is a function of the end point positions only.*

PROOF. Suppose that (2) reduces to a zero identity, then on writing out the equation in full we have

$$\frac{\partial F}{\partial y}-\frac{\partial^2 F}{\partial x\partial y'}-\frac{\partial^2 F}{\partial y\partial y'}\frac{dy}{dx}-\frac{\partial^2 F}{\partial y'^2}\frac{d^2y}{dx^2} = 0. \tag{4}$$

Since the factor d^2y/dx^2 occurs only in the last term on the left-hand side, if (4) is an identity then $\partial^2F/\partial y'^2 = 0$, i.e. F can contain y' only to the first degree. Hence F must be of the form

$$F = py'+q, \tag{5}$$

where p and q are functions of x and y only. On substituting this in (4) we get

$$\frac{\partial p}{\partial y}y'+\frac{\partial q}{\partial y}-\frac{\partial p}{\partial x}-\frac{\partial p}{\partial y}y' = 0. \tag{6}$$

Hence $\partial q/\partial y = \partial p/\partial x$ and so $p\,dy+q\,dx$ must be a perfect differential.

Let $$p\,dy+q\,dx = df(x, y), \tag{7}$$

then from (5) we obtain immediately

$$\int F(x,y,y')\,dx = f(x,y). \tag{8}$$

It is evident from (8) that the value of the integral must be a function of the end points only.

EXAMPLE 14. In example 6, § 1.11, find the brachistochrone if the particle is projected from A with the velocity $(2gh)^{\frac{1}{2}}$.

EXAMPLE 15. A plane curve is a free orbit under one central force and a brachistochrone under another central force to the same point. v, v_1 are the velocities of the respective particles at the same point (although not necessarily at the same time). Prove that vv_1 is constant.

EXAMPLE 16. Prove that an ellipse is a brachistochrone for a particle moving under a central force to one focus varying inversely as the square of the distance from the other focus.

CHAPTER II

THE SECOND VARIATION

2.1. Introduction

IN this chapter we shall deal with the second variation, see § 1.3, equation (11). The ideas upon which our study is based are much the same as those used in the maximum and minimum theory of functions of one variable. If $f'(a) = 0$, then we can deduce at most that $f(x)$ is stationary at $x = a$. Further information about the nature of $f(a)$ requires a study of the sign of $f''(a)$. If this is positive (negative), then $f(x)$ has a minimum (maximum) value at $x = a$, and if $f''(a) = 0$, then derivatives of order higher than the second must be considered.

The corresponding study in the calculus of variations is necessarily much more complex than this.

2.2. The second variation

Recapitulating the investigations of the previous chapter, especially those of §§ 1.3 and 1.4, we have the following result:

Let $y = s(x)$ be the path of integration for which the integral

$$I = \int_a^b F(x, y, y')\, dx, \tag{1}$$

is stationary and let $y = s(x)+\epsilon t(x)$ be the path when a weak variation is made in y. If I_s and $I_s+\delta I_s$ are the corresponding values of I, then

$$\delta I_s = I_2+O(\epsilon^3), \tag{2}$$

where I_2, the second variation, is given by

$$I_2 = \frac{\epsilon^2}{2!}\int_a^b \left\{t^2\frac{\partial^2 F}{\partial s^2}+2tt'\frac{\partial^2 F}{\partial s\partial s'}+t'^2\frac{\partial^2 F}{\partial s'^2}\right\} dx. \tag{3}$$

Here dashes denote differentiation with respect to x and, for brevity, $t(x)$ has been replaced by t.

In order that I_s should be a maximum the sign of δI_s must be negative independently of the choice of $t(x)$, for all sufficiently small ϵ. This requires that the integral of (3) should be negative independently of the choice of $t(x)$. Similar remarks, but with changed sign, apply to the case when I_s is a minimum.

By means of an ingenious artifice, due to Jacobi, this integral can be put into a form which shows that its sign depends almost entirely upon the sign of $\partial^2F/\partial s'^2$. The transformation will be divided into two stages, each of which will be dealt with by a simple lemma. The second of these will introduce the Jacobi accessory equation.

For simplicity the following notation will be adopted:

$$\frac{\partial F}{\partial s} = F_0,\ \frac{\partial F}{\partial s'} = F_1,\ \frac{\partial^2 F}{\partial s^2} = F_{00},\ \frac{\partial^2 F}{\partial s \partial s'} = F_{01},\ \frac{\partial^2 F}{\partial s'^2} = F_{11}, \quad (4)$$

so that we have

$$I_2 = \frac{\epsilon^2}{2!} \int\limits_a^b (t^2F_{00}+2tt'F_{01}+t'^2F_{11})\,dx. \quad (5)$$

2.3. Lemma 1

If $t(a) = t(b) = 0$, then

$$\int\limits_a^b (t^2F_{00}+2tt'F_{01}+t'^2F_{11})\,dx$$
$$= \int\limits_a^b \left\{(t^2F_{00}-t^2\frac{d}{dx}(F_{01})-t\frac{d}{dx}(t'F_{11})\right\}dx. \quad (1)$$

To prove this it is evident that the right-hand side is equal to

$$\int\limits_a^b \left\{t^2F_{00}+tt'F_{01}-t\frac{d}{dx}(tF_{01}+t'F_{11})\right\}dx.$$

On integrating the negative terms by parts this becomes equal to

$$\int\limits_a^b (t^2F_{00}+tt'F_{01})\,dx-t(tF_{01}+t'F_{11})_a^b+\int\limits_a^b t'(tF_{01}+t'F_{11})\,dx. \quad (2)$$

But, by hypothesis, $t(a) = t(b) = 0$ and so $t(tF_{01}+t'F_{11})_a^b = 0$. Hence expression (2) reduces to the left-hand side of (1) and the truth of the lemma is established.

The conditions $t(a) = t(b) = 0$ are those stipulated in § 1.3 in order that $t(x)$ should vanish at the end points of the curve APB (Fig. I. 1).

2.4. Lemma 2. Jacobi's accessory equation

On solving the characteristic equation (3), § 1.4, the equation of the extremal $y = s(x)$ which passes through the given points A and B can be determined.

Thus the quantities F_{00}, F_{01}, F_{11}, and $\frac{d}{dx}(F_{01})$ can all be expressed in terms of x and the differential equation

$$\left\{F_{00} - \frac{d}{dx}(F_{01})\right\} u - \frac{d}{dx}\left(F_{11}\frac{du}{dx}\right) = 0 \tag{1}$$

can then be solved for u as a function of x. This is an ordinary linear differential equation of the second order and is known as the subsidiary or Jacobi's equation or, more frequently, as the accessory equation.

On taking x to be the independent and t $(= t(x))$ the dependent variable in the integral of I_2, (5), § 2.2, it is easily seen that (1) is the Euler equation for minimizing I_2 with t replaced by u.

Lemma 2. If $t(a) = t(b) = 0$ and u is a solution of equation (1), then

$$I_2 = \frac{\epsilon^2}{2!}\int_a^b F_{11}\left(t' - t\frac{u'}{u}\right)^2 dx, \tag{2}$$

where $t(x)$ is denoted by t and, as usual, dashes or primes denote differentiation.

Proof. From lemma 1, we have

$$I_2 = \frac{\epsilon^2}{2!}\int_a^b \left\{t^2 F_{00} - t^2\frac{d}{dx}(F_{01}) - t\frac{d}{dx}(t'F_{11})\right\} dx, \tag{3}$$

and from equation (1) this can evidently be written in the form

$$I_2 = \frac{\epsilon^2}{2!}\int_a^b \frac{t}{u}\left\{t\frac{d}{dx}\left(F_{11}\frac{du}{dx}\right) - u\frac{d}{dx}\left(F_{11}\frac{dt}{dx}\right)\right\} dx. \tag{4}$$

Now by straightforward differentiation it is easily shown that

$$t\frac{d}{dx}\left(F_{11}\frac{du}{dx}\right) - u\frac{d}{dx}\left(F_{11}\frac{dt}{dx}\right) = \frac{d}{dx}\left\{F_{11}\left(t\frac{du}{dx} - u\frac{dt}{dx}\right)\right\}, \tag{5}$$

so that equation (4) can be written in the form

$$I_2 = \frac{\epsilon^2}{2!}\int\limits_a^b \frac{t}{u}\frac{d}{dx}\{F_{11}(tu'-ut')\}\,dx. \tag{6}$$

On integrating by parts we have

$$I_2 = \frac{\epsilon^2}{2!}\left\{\frac{t}{u}F_{11}(tu'-ut')_a^b - \int\limits_a^b F_{11}(tu'-ut')\frac{d}{dx}\left(\frac{t}{u}\right)dx\right\}. \tag{7}$$

But, by hypothesis, $t(a) = t(b) = 0$ and so the terms not inside the integral vanish. On evaluating $\frac{d}{dx}\left(\frac{t}{u}\right)$ equation (7) then reduces to (2).

One point remains to be considered. One of the terms on the right-hand side of (7) is $(t^2F_{11}u'/u)_a^b$ and it may be possible that $u(a)$ or $u(b)$ vanishes. Towards the end of this chapter, in § 2.18, some general properties of the solutions of equation (1) will be established, and among them it will be shown that $u(x)$ cannot have a double root. Anticipating this result, it follows that even if $u(x)$ vanishes at either or both of the values $x = a$ and $x = b$, both $t^2(a)/u(a)$ and $t^2(b)/u(b)$ still vanish since $t(a) = t(b) = 0$ by hypothesis.

Since all the squared factors on the right-hand side of (2) must be positive or zero it is evident that the sign of I_2 is more easily determined from equation (2) than from its definition in equation (3), § 2.2.

2.5. Simple criteria for maxima and minima of I. The Legendre test

The function $\{t'-t(u'/u)\}^2$ must always be positive or zero. Consider first the case when it is always greater than zero except at a finite number of isolated points of the extremal arc AB. The difficult but important case when $t(x)$ can be chosen so that $\{t'-t(u'/u)\}$ vanishes at all points of the arc AB will be postponed to the next paragraph.

From equation (2), § 2.4, it is evident that if $\{t'-t(u'/u)\} \neq 0$ and F_{11} has constant sign for all points of the extremal arc AB, then I_2 must have a sign which is independent of the choice of either ϵ or $t(x)$. Now in the stationary case we have

$$\delta I_s = I_2 + O(\epsilon^3),$$

(2), § 2.2, and so the value of I will be a maximum if $\partial^2 F/\partial s'^2$ is negative (a minimum if it is positive) at *all* points of the extremal arc AB. This is essentially the Legendre test anticipated in theorem 4, § 1.5.

If, however, F_{11} does not keep its sign constant at all points of the extremal arc AB, then it is easy to show that I_s, the corresponding stationary value of I, is neither a maximum nor a minimum. Suppose that F_{11} is positive for values of x lying between a and c, where $a < c < b$, and negative for values of x lying between c and b. Then if $t(x)$ is chosen so as to vanish in the range a to c and not to vanish in the range c to b, we see from (2), § 2.4 that I_2 is negative. On the other hand, if $t(x)$ is chosen so as not to vanish in the range a to c and to vanish in the range c to b, then I_2 is positive. Hence if $\partial^2 F/\partial s'^2$ does not keep constant sign for all points of the arc AB along the extremal, I_2 can be made either positive or negative by suitable choice of $t(x)$ and the integral $\int_a^b F(x, y, y')\,dx$ taken along $y = s(x)$ can be neither a maximum nor a minimum.

2.6. Conjugate points (kinetic foci)

The possibility that $t(x)$ can be chosen so that

$$t' - t\frac{u'}{u} = 0 \tag{1}$$

at all points of the extremal arc AB must now be dealt with. The function u is known, since it is a solution of the accessory equation (1), § 2.4, and equation (1) may therefore be regarded as a differential equation for $t(x)$. The solution of this equation is

$$t(x) = \alpha u(x), \tag{2}$$

where α is an arbitrary constant (assumed other than zero).

Along an extremal the first variation I_1 vanishes by virtue of the characteristic equation, §§ 1.3 and 1.4. If in addition $t(x)$ is chosen so as to satisfy equation (2), then by (2), § 2.4, the second variation I_2 will also vanish. The sign of I_s will then depend upon the third variation I_3, where

$$I_3 = \frac{\epsilon^3}{3!}\int\limits_a^b \left\{t^3\frac{\partial^3 F}{\partial s^3}+3t^2t'\frac{\partial^3 F}{\partial s^2\partial s'}+3tt'^2\frac{\partial^3 F}{\partial s\partial s'^2}+t'^3\frac{\partial^3 F}{\partial s'^3}\right\}\,dx. \quad (3)$$

Since the sign of I_3 depends upon that of ϵ there can be no maximum or minimum values of I unless I_3 vanishes, in which case the sign of δI_s will depend upon that of I_4, the fourth variation. Thus if $t(x)$ can be chosen so as to satisfy (2) we are faced with the problem of establishing conditions which ensure (i) that I_3 vanishes and (ii) that I_4 has a constant sign independent of the choice of ϵ and $t(x)$. In order to avoid these difficult problems we proceed to establish a test (Jacobi's test) which, if satisfied, makes it impossible to choose $t(x)$ so as to satisfy (2). Thus when Jacobi's test is satisfied the results of § 2.5 are valid.

The test depends upon the following idea. It has been stipulated in § 1.3 that $t(x)$ vanishes both when $x = a$ and $x = b$. If, therefore, $u(x)$ can be made to vanish at only one of these points, say at $x = a$ and *not* at $x = b$, then $t(x)$ cannot be chosen to satisfy equations (1) and (2) of this paragraph. This leads us to the following definition:

Definition of conjugate points or kinetic foci. Let $u(x)$, a solution of the accessory equation (1), § 2.4, be chosen so as to vanish at the point A, where $x = a$. Then all other points on the extremal $y = s(x)$ at which $u(x)$ vanishes are known as points conjugate to A. In dynamics they are generally known as the kinetic foci of A.

From (1), § 2.4, $u(x)$ is the solution of a second-order differential equation and is therefore of the form

$$u(x) = \beta_1 u_1(x)+\beta_2 u_2(x), \quad (4)$$

where $u_1(x)$ and $u_2(x)$ are independent solutions of (1), § 2.4, and β_1 and β_2 are arbitrary constants. If $u(x) = 0$ when $x = a$, then from (4) we have

$$\frac{u_1(a)}{u_2(a)} = -\frac{\beta_2}{\beta_1}. \quad (5)$$

If x is the abscissa of any point conjugate to A, then $u(x) = 0$ Hence from (4) and (5)

$$\frac{u_1(x)}{u_2(x)} = -\frac{\beta_2}{\beta_1} = \frac{u_1(a)}{u_2(a)}. \tag{6}$$

This is the equation for the abscissae of all points conjugate to the point A (where $x = a$). It shows that the ratio $u_1(x)/u_2(x)$ is the same for all conjugate points.

Proceeding with Jacobi's test, suppose that the point B, where $x = b$, is *not* situated at a point conjugate to A. Then $u(b)$ cannot vanish. Therefore, since we must have $t(b) = 0$, $t(x)$ cannot be chosen to be proportional to $u(x)$ for the whole arc AB. In other words, equation (1) cannot hold if B is not conjugate to A.

Let A' be the first point conjugate to A when moving along the extremal in the direction from A to B. There are three possible cases, as follows:

Case (i) B lies between A and A'.

Case (ii) B coincides with A'.

Case (iii) B lies beyond A'.

In all three cases $u(x)$ vanishes at A, but in case (i) it cannot vanish again at any point of arc AB, in case (ii) it vanishes again at B, and in case (iii) it vanishes again at some point of the arc AB lying between A and B.

In case (i) we have $u(a) = t(a) = 0$, $u(b) \neq 0$, $t(b) = 0$. Hence $u(x)$ cannot be proportional to $t(x)$ at all points of the extremal arc AB. Therefore we must have $\left(t' - t\dfrac{u'}{u}\right)^2 > 0$ at all points of AB, except, possibly, at a finite number of points where $t(x)$ and $t'(x)$ vanish simultaneously.

Theorem 6. *Let $y = s(x)$ be the equation of the extremal through the points A and B for which the integral $I = \int_a^b F(x, y, y')\,dx$ is stationary. Let A and A' be two adjacent conjugate points (kinetic foci) on the curve. If* (i) *B lies between A and A' and* (ii) *$\partial^2 F/\partial s'^2$ has constant sign for all points of the arc AB, then for weak* *variations I is a maximum when $\partial^2 F/\partial s'^2$ is negative and a minimum when it is positive.*

This result is due to Jacobi (1837).

2.7. Case when B does not lie between A and its nearest conjugate

In case (ii), § 2.6, where B coincides with A we have

$$u(a) = t(a) = 0 \quad \text{and} \quad u(b) = t(b) = 0.$$

Hence $t(x)$ can be chosen to be proportional to $u(x)$ at all points of arc AB. In case (iii), where B lies beyond A', $t(x)$ can be chosen so as to be proportional to $u(x)$ at all points of the arc AA' and zero at all points of the arc $A'B$. In such circumstances, as explained in the previous section, the first and second variations both vanish and the sign of δI_s depends upon that of I_3, the third variation, given by (3), § 2.6. For maxima or minima of I the third variation I_3 must also vanish and the sign of δI_s will then depend upon that of I_4. Such problems are too complex to be dealt with in this book and we shall therefore confine ourselves entirely to those cases in which there are sufficient conditions to ensure the truth of theorem 6.

These problems have been investigated by various mathematicians, and Weierstrass has proved that if B lies beyond A' then the stationary values of I are neither maxima nor minima.†

2.8. The accessory equation

In order to apply the theory developed so far two differential equations must be solved, the characteristic and the accessory equations. The following investigations will show that the solutions of the accessory equation can be derived from those of the characteristic equation by differentiation.

Replacing s by y and using the terminology of § 2.2, the characteristic and accessory equations take the forms

$$F_0 - \frac{dF_1}{dx} = 0 \tag{1}$$

and

$$\left(F_{00} - \frac{dF_{01}}{dx}\right)u - \frac{d}{dx}\left(F_{11}\frac{du}{dx}\right) = 0. \tag{2}$$

† Forsyth, *Calculus of Variations*, p. 111 et seq.

Equation (1) is a second-order differential equation for y and its solution is of the form

$$y = s(x, c_1, c_2), \tag{3}$$

where c_1 and c_2 are arbitrary constants. We now show that $\partial y/\partial c_1$ and $\partial y/\partial c_2$ are solutions of (2).

Both F_0 and F_1 contain y and y', and these in turn contain c_1. Hence, on differentiating (1) partially with respect to c_1, we have

$$\frac{\partial F_0}{\partial y}\frac{\partial y}{\partial c_1}+\frac{\partial F_0}{\partial y'}\frac{\partial y'}{\partial c_1}-\frac{d}{dx}\left(\frac{\partial F_1}{\partial y}\frac{\partial y}{\partial c_1}+\frac{\partial F_1}{\partial y'}\frac{\partial y'}{\partial c_1}\right) = 0. \tag{4}$$

We now write

$$u = \frac{\partial y}{\partial c_1} \quad \text{and} \quad u' = \frac{\partial y'}{\partial c_1} \tag{5}$$

and make use once again of the terminology of § 2.2, with s replaced by y. Equation (4) then easily reduces to (2). Thus $\partial y/\partial c_1$ is a solution of (2) and similar arguments show that $\partial y/\partial c_2$ is also a solution.

From these results we shall make two important deductions, the first in the next section and the second in § 2.12.

2.9. A property of conjugate points

Let k be a constant, $u(x)$ a solution of the accessory equation (2), § 2.8, and $s(x, c_1, c_2)$ a solution of the characteristic equation (1), § 2.8. Then the curves

$$y = s(x, c_1, c_2) \quad \text{and} \quad y = s(x, c_1, c_2)+ku(x)$$

intersect at points given by $u(x) = 0$. But if $u(x)$ vanishes at A it also vanishes at all points conjugate to A (§ 2.6). Hence if these curves intersect at A they also intersect at the conjugates of A.

Now consider two adjacent extremals

$$y = s(x, c_1, c_2) \tag{1}$$

and

$$y = s(x, c_1+\delta c_1, c_2), \tag{2}$$

where δc_1 is small. Neglecting $O(\delta c_1)^2$ (2) may be written in the form

$$y = s(x, c_1, c_2)+\frac{\partial y}{\partial c_1}\delta c_1. \tag{3}$$

But in the previous section it was shown that $\partial y/\partial c_1$ is a solution of (2), § 2.8, the accessory equation, and therefore, by § 2.6, if it vanishes at A it also vanishes at the conjugates of A. Hence if (1) and (2) intersect at A they must also intersect at all its conjugates.

Consequently we may conclude that conjugates of A are the limiting point of intersection of neighbouring extremals passing through A.

As a corollary the following result can be deduced from the geometrical theory of envelopes. If the family of extremals passing through A has an envelope E, then the points of contact of E and the extremal through AB are the conjugates of A.

Some of the examples dealt with in the first chapter can now be completed. Consider the problem of finding the shortest distance between two points A and B, example 1, § 1.6. The extremals are straight lines and the Legendre test shows that a minimum is possible if the length AB is sufficiently small. To apply the ideas of this chapter we must first determine the conjugates of A either by (6), § 2.6, or by finding the points of intersection of AB and a neighbouring extremal. The latter method is easier in this case since straight lines through A cannot intersect again at any other point. Hence there is no point conjugate to A and the straight line is the shortest distance between A and B whatever the length of AB.

Two further illustrations are given by the problems of geodesics on a sphere, § 1.10, and of optical paths in minimum time, § 1.9.

In § 1.10 it is shown that geodesics on a sphere are arcs of great circles. If AA' is a diameter of a sphere, then all great circles on the sphere passing through A are concurrent at A'. A' therefore is the conjugate of A. The Legendre test of § 1.10 shows that if the arc of the great circle between A and B is sufficiently small then it may be a minimum. The Jacobi test shows in addition that if B lies between A and its diametrically opposite point A', i.e. if AB subtends an angle less than π at the centre of the sphere, then the length of the great circle arc AB is a minimum for all curves on the sphere joining A to B.

If AB subtends an angle greater than π at the centre, then the great circle arc AB is stationary but is neither a maximum nor a minimum.

In the case of optical paths considered in § 1.9 the conjugate of A can be found easily if we confine ourselves to the case of the symmetrical optical instrument. If A is not too distant from the axis of the instrument, then all rays through A making small angles with the axis are concurrent at a point A' known as the optical image of A in the instrument. Hence the conjugate of A is its optical image, and if B lies between A and its optical image A', the path of a ray of light from A to B is such that the time from A to B is a minimum.

2.10. Principle of least action

In example 4, § 1.8, it has been proved, by means of the principle of least action, that the trajectory of a particle moving in the earth's gravitational field is a parabola.

A particle is projected from a given point A with a given initial speed u. If we confine ourselves to one vertical plane through A and vary the direction of projection, keeping u constant, a family of parabolas will be generated. Then it is known that this family of parabolas has a parabolic envelope whose focus is at A and whose principal diameter is the vertical through A. For completeness we give a proof of this statement.

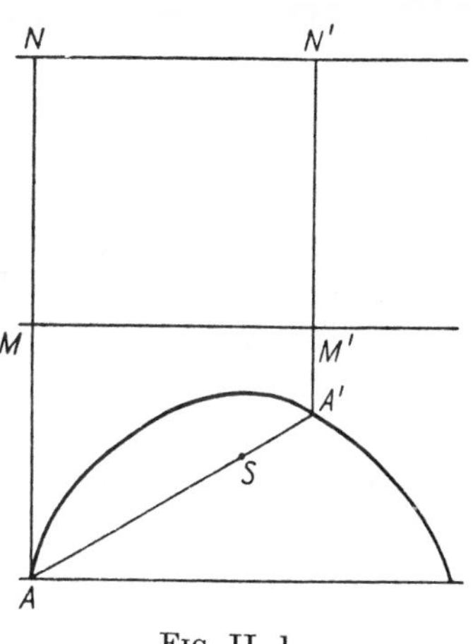

FIG. II. 1.

Fig. II. 1 shows a parabolic trajectory through A having S as its focus and MM' as its directrix. By elementary theory if the angle of projection with the horizontal is α the equation of the trajectory is

$$2yu^2\cos^2\alpha = 2xu^2\cos\alpha\sin\alpha - gx^2. \tag{1}$$

It is easy to show that the height of the directrix of (1) above A

is $u^2/2g$. Let AS meet the parabola again at A', let M and M' be the feet of the perpendiculars from A and A' respectively to the directrix, and let $MM'N'N$ be a rectangle with

$$MN = M'N' = u^2/2g.$$

Then

$$AA' = AS+SA' = AM+A'M' = M'N'+A'M' = A'N'.$$

Hence the locus of A' is a parabola with A as focus and NN' as directrix.

Again the tangent to the trajectory at A' bisects the angle $SA'M'$ and the tangent to the locus of A' bisects the angle $AA'N'$, hence the tangents at A' to the trajectory and to the locus of A' are the same. Summing up these results it is clear that (i) the envelope of the family of trajectories is the locus of A' (i.e. a parabola with A as focus), and (ii) that A' is the point of contact with the envelope. Hence from § 2.9 A' is the conjugate of A.

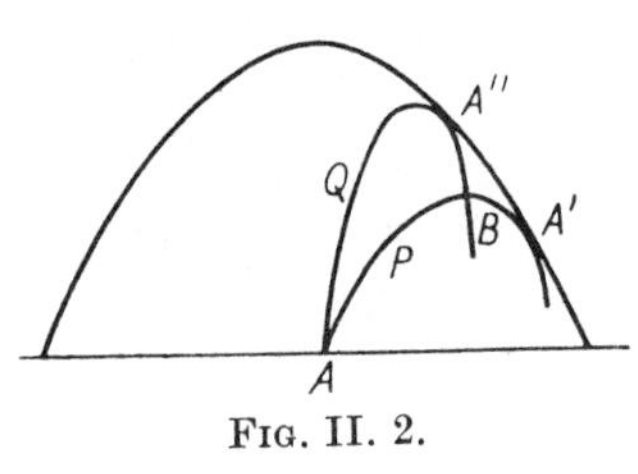

Fig. II. 2.

Now it is known that if u, the speed of projection, is sufficiently large, then there are two possible trajectories through the two points A and B. This is shown in Fig. II. 2, where APB and AQB are the two possible trajectories. Let the parabolic envelope touch APB at A' and AQB at A''. Then A' and A'' are conjugates of A. Also, Fig. II. 2, it is evident that if B lies inside the arc AA' it must lie outside the arc AA''. Now the results of example 4, § 1.8, tell us that the action, $2\int_A^B T\,dt$, taken along either of the arcs APB or AQB is stationary. The Jacobi test tells us in addition that for arc APB the action is a minimum and for arc AQB the action is neither a maximum nor a minimum.

2.11. The catenary

We now consider example 3, § 1.7, where it is shown that a uniform flexible chain hangs under gravity in the form of a catenary. If the pegs from which the chain hangs are on the same horizontal level and if their coordinates are $(\pm h, k)$, then there will be two possible catenaries through A and B provided that $k > h \tan \alpha$, where $\alpha = 56^\circ\, 28'$ approximately. It is also shown that the extremals are catenaries whose equations are $y = c \cosh(x/c)$. These all have the x-axis as a common directrix.

It is easy to prove that these catenaries touch the lines $y = \pm x \tan \alpha$ $(\alpha = 56^\circ\, 28')$ for all values of c. This is illustrated in Fig. II. 3 which shows the two possible catenaries through A and B and their points of contact with these lines at A', B' and A'', B''.

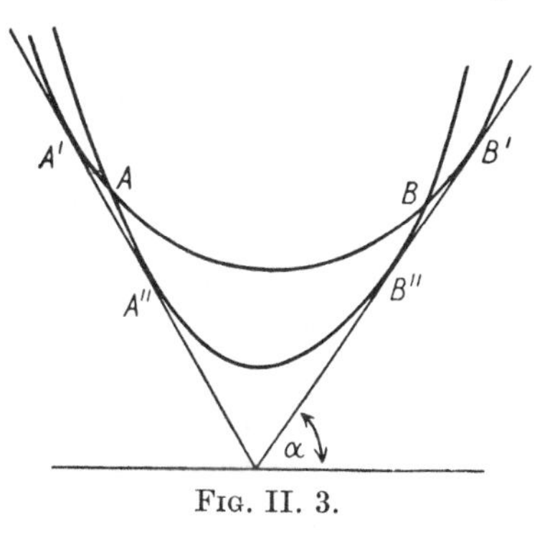

FIG. II. 3.

It is evident from the diagram that for the upper catenary, which corresponds to the larger value of c, B lies between A and its conjugate point B'. Therefore for this catenary the potential energy is a minimum and we have stable equilibrium. But for the lower catenary, which corresponds to the smaller value of c, B lies beyond A'', the first conjugate of A. Therefore the potential energy of this catenary, although stationary, is neither a maximum nor a minimum. Thus for the lower catenary the equilibrium is unstable.

This investigation applies also to example 7, § 1.12, where minimal surfaces are discussed.

2.12. Analytical methods for finding conjugate points

In the previous examples the positions of the points conjugate to A have been found either by physical considerations or because the envelope of the extremals through A can be found without much difficulty. When the conjugate points cannot be found easily by such means we must revert to the ideas of § 2.6. There we proved the following result:

If a is the abscissa of the point A and x that of a point conjugate to A, then

$$\frac{u_1(x)}{u_2(x)} = \frac{u_1(a)}{u_2(a)}, \tag{1}$$

where $u_1(x)$ and $u_2(x)$ are independent solutions of the accessory equation, (1), § 2.4.

The characteristic equation (8), § 1.4, is a linear differential equation of the second order having a solution of the form

$$y = s(x, c_1, c_2), \tag{2}$$

where c_1 and c_2 are arbitrary constants. Now it has been shown in § 2.8 that $\partial y/\partial c_1$ and $\partial y/\partial c_2$ are both solutions of the accessory equation, (1), § 2.4. It follows that (1) can be replaced by the more convenient result

$$\frac{\partial y}{\partial c_1}\Big/\frac{\partial y}{\partial c_2} = \left(\frac{\partial y}{\partial c_1}\Big/\frac{\partial y}{\partial c_2}\right)_{(x=a)} \tag{3}$$

The solutions of this equation are the abscissae of points conjugate to A (abscissa a).

In some applications it is more convenient to use polar coordinates than Cartesians. If the vectorial angle is taken as the independent variable, then a and x in (3) denote, respectively, the vectorial angles of A and a conjugate point.

We now apply this analysis to some of the problems of Chapter I.

2.13. Conjugate points on the catenary

In § 1.7 we dealt with the problem of the flexible chain by using the principle that a statical system is in stable equilibrium when its potential energy is a minimum. It was shown in § 1.7 that the curve in which the chain must hang, in order to give a stationary value of the potential, is a catenary. Its equation is

$$y = c\cosh\left(\frac{x+b}{c}\right), \tag{1}$$

where b and c are arbitrary constants and the x-axis is the directrix.

In § 2.11 we dealt with a special case of this result when the two pegs over which the chain is suspended have coordinates of the form $(\pm h, k)$. Here we shall deal with the problem in greater generality.

In order to use the results of § 2.12 we must differentiate the extremal (1) partially with respect to c and b. We have

$$\left.\begin{aligned}\frac{\partial y}{\partial c} &= \cosh\left(\frac{x+b}{c}\right)-\left(\frac{x+b}{c}\right)\sinh\left(\frac{x+b}{c}\right)\\ \frac{\partial y}{\partial b} &= \sinh\left(\frac{x+b}{c}\right)\end{aligned}\right\}. \tag{2}$$

From (3), § 2.12, the equation for the abscissae of points conjugate to A becomes

$$\coth\left(\frac{x+b}{c}\right)-\left(\frac{x+b}{c}\right) = \coth\left(\frac{a+b}{c}\right)-\left(\frac{a+b}{c}\right), \tag{3}$$

where a is the abscissa of the point A.

This equation has a simple geometrical interpretation. In Fig. II. 4 the catenary (1) is shown with its vertex at V and with U as the foot of the perpendicular from V to the directrix (which lies along the x-axis). Since y in (1) has a minimum when $x = -b$, it follows that $UO = b$, where O is the origin. P is any point on the curve, with abscissa equal to x, M is the foot of the perpendicular from P to the x-axis, i.e. $OM = x$, and T is the point of intersection of the tangent to the catenary at P and the x-axis. We have

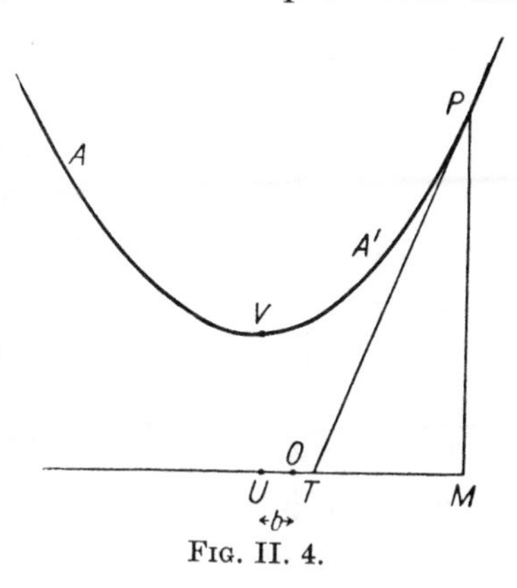

FIG. II. 4.

$$\begin{aligned}TM &= MP/\tan MTP\\ &= c\cosh\left(\frac{x+b}{c}\right)\Big/\sinh\left(\frac{x+b}{c}\right).\end{aligned} \tag{4}$$

Hence $$\frac{UT}{c} = \frac{UM-TM}{c} = \left(\frac{x+b}{c}\right)-\coth\left(\frac{x+b}{c}\right). \tag{5}$$

If the tangents to the catenary at A and at its conjugate A' intersect the x-axis at N and N' respectively, then from (3) and (5) we have

$$UN' = UN. \tag{6}$$

Therefore the tangents to the catenary at A and its conjugate point A' must intersect on the directrix.

As z varies from $-\infty$ to ∞, $\coth z - z$ decreases steadily from ∞ to $-\infty$ twice. Thus $\coth z = z+a$ can be satisfied by two real values of z only. Hence (3), treated as an equation in x, can have only one real root other than $x = a$ and therefore only two real tangents can be drawn from a point on the directrix to the catenary. Hence there can be only one point conjugate to A. In addition, if A' is conjugate to A, then, conversely, A is conjugate to A'.

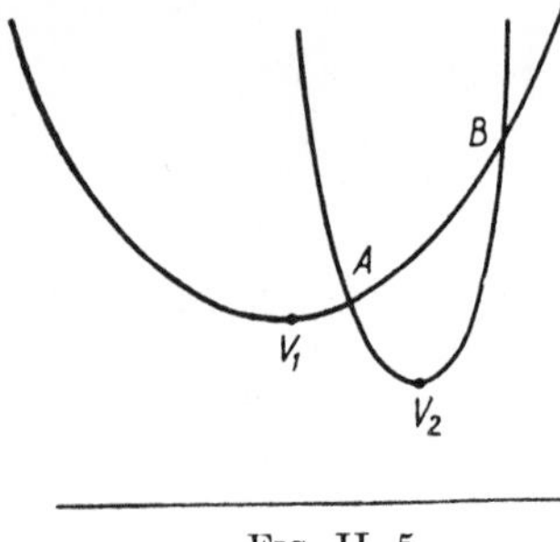

FIG. II. 5.

The problem of the flexible chain hanging over two pegs can now be completed. In general only two catenaries can have a common directrix and pass through two given points A and B.† In Fig. II. 5 the two catenaries through A and B are distinguished from each other by their vertices V_1 and V_2. To apply the Jacobi test we find the conjugates of A for each of the two catenaries, by the tangent property proved above. It is seen that A lies between B and its conjugate only for the upper of the two catenaries, i.e. for the one with the larger value of c. Hence only for the upper catenary is the potential energy a minimum, and the equilibrium therefore stable.

2.14. Conjugate points on a parabolic trajectory

The conjugate points of the parabolic trajectory, discussed in § 1.8, have already been obtained geometrically in § 2.10. In this section we shall again find the conjugate points by using the analytical theory of § 2.12. The two methods can thus be contrasted.

The equation of the family of trajectories, (8), § 1.8, is

$$2gy = u^2 - c^2 - (g^2/c^2)(x-d)^2, \tag{1}$$

where u, the speed of projection, is a known constant and c

† For the proof of this statement see Forsyth, *Calculus of Variations*, p. 98 et seq.

and d are arbitrary constants. In equation (3), § 2.12, we write $c_1 = c$, $c_2 = d$ and on taking the origin at A we have in addition $a = 0$. The equation for the abscissa of A', the conjugate of A, then becomes

$$\frac{-2c+(2g^2/c^3)(x-d)^2}{(2g^2/c^2)(x-d)} = \frac{-2c+(2g^2/c^3)d^2}{-(2g^2/c^2)d}, \tag{2}$$

which simplifies easily to

$$(x-d)d = c^4/g^2. \tag{3}$$

In Fig. II. 6 A is the origin, S is the focus of the trajectory under consideration, V is the vertex, and U is the foot of the perpendicular from V to the x-axis. From (1) y is a maximum when $x = d$, hence $AU = d$. Therefore, if M is the foot of the perpendicular from A' to UV, equation (3) is equivalent to

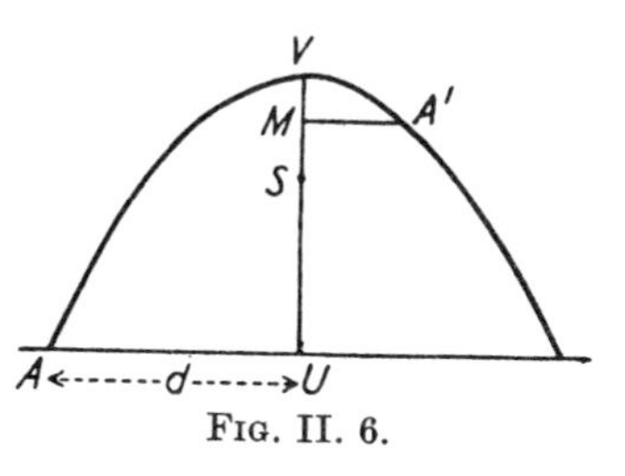

FIG. II. 6.

$$MA'.AU = c^4/g^2. \tag{4}$$

But from (1) the latus rectum of the parabola is $2c^2/g$. Hence from well-known and easily proved properties of the parabola, equation (4) is the condition that the chord AA' should pass through S, the focus of parabola (1). This is in complete agreement with the result obtained geometrically in § 2.10.

2.15. Geodesics on spheres

This problem has been dealt with in § 1.10. The equation of the geodesics, (7), § 1.10, is

$$\phi+\beta = \cos^{-1}\left(\frac{\tan\alpha}{\tan\theta}\right), \tag{1}$$

where θ is the co-latitude, ϕ is the longitude, and α and β are arbitrary constants. In § 2.9 it was proved, by a geometrical method, that conjugate points on a sphere are at the opposite ends of a diameter; we shall obtain the same result here by means of the analytical methods of § 2.12.

On taking θ as the independent and ϕ as the dependent variable, equation (3), § 2.12, becomes

$$\frac{\partial\phi}{\partial\alpha}\bigg/\frac{\partial\phi}{\partial\beta} = \left(\frac{\partial\phi}{\partial\alpha}\bigg/\frac{\partial\phi}{\partial\beta}\right)_{\theta=\theta_0}, \tag{2}$$

where θ_0 and θ are respectively the co-latitudes of A and A' (the conjugate of A).

Applying this result to (1) above we have

$$\frac{\sec^2\alpha}{\tan\theta}\frac{1}{\left(1-\dfrac{\tan^2\alpha}{\tan^2\theta}\right)^{\frac{1}{2}}} = \frac{\sec^2\alpha}{\tan\theta_0}\frac{1}{\left(1-\dfrac{\tan^2\alpha}{\tan^2\theta_0}\right)^{\frac{1}{2}}}, \tag{3}$$

whose solution is $\theta = \theta_0+m\pi$, when m is an integer. It is evident from (1) that two points on a great circle whose co-latitudes differ by a multiple of π must have the same longitude. Hence if the coordinates of A are (θ_0, ϕ_0), those of its conjugate A' are $(\theta_0+\pi, \phi_0)$, and so A and A' are diametrically opposite points, in agreement with the result of § 2.9.

2.16. Orbits under inverse square law of attraction

The orbit of a particle of mass m, attracted to a fixed centre of force O by the Newtonian law $m\mu/r^2$ where μ is constant and r is its distance from O, has been investigated in § 1.13. The equation of the orbit in polar coordinates is (9), § 1.13,

$$\frac{c}{r} = 1+\left(1-\frac{c}{a}\right)^{\frac{1}{2}}\cos(\theta+\beta). \tag{1}$$

Here β and c are arbitrary constants and a is a constant of known value, depending upon the initial speed of the particle. This is the equation of a conic, and if e is the eccentricity we have $c = a(1-e^2)$. Equation (1) can then be written

$$r = \frac{a(1-e^2)}{1+e\cos(\theta+\beta)}. \tag{2}$$

In order to apply the methods of § 2.12 we must first put equation (3), § 2.12, in the form appropriate to the variables and constants of (2) above. We write $c_1 = e$, $c_2 = \beta$, and take θ to be the independent and r the dependent variable. We choose OA to be the initial line and then β is equal to the angle between OA

and the major axis of the orbit. Equation (3), § 2.12, takes the form

$$\left(\frac{\partial r}{\partial e}\Big/\frac{\partial r}{\partial \beta}\right) = \left(\frac{\partial r}{\partial e}\Big/\frac{\partial r}{\partial \beta}\right)_{\theta=0}. \tag{3}$$

On applying this to (2) above we obtain

$$\frac{2e+(1+e^2)\cos(\theta+\beta)}{\sin(\theta+\beta)} = \frac{2e+(1+e^2)\cos\beta}{\sin\beta}. \tag{4}$$

Multiplying out and dividing through by $\sin\frac{1}{2}\theta$, whose zeros correspond to the point A only, we can then solve for $\tan\frac{1}{2}\theta$. The final result is

$$\theta = 2\tan^{-1}\left\{\frac{1+2e\cos\beta+e^2}{2e\sin\beta}\right\}, \tag{5}$$

which gives the vectorial angle of A', the conjugate of A. Denoting the right-hand side of (5) by α, where $0 < \alpha < 2\pi$, it follows that if $0 < \widehat{AOB} < \alpha$ then the action $2\int T\,dt$ is a minimum, and if $\widehat{AOB} > \alpha$ then the action is stationary but is neither a maximum nor a minimum.

When the conic reduces to a circle, $e = 0$, the points A and A' are at opposite ends of a diameter and when the conic is a parabola, $e = 1$, A' is at infinity.

2.17. Orbit of a particle attracted by a force $m\mu r$

In example 9, § 1.14, we found the orbit of a particle attracted towards a fixed centre of force O by a force $m\mu r$, where m is the mass, μ is a constant, and r is its distance from O. The polar equation of the orbit, (4), § 1.14, obtained by the principle of least action, is:

$$\frac{c^2}{r^2} = \left(\frac{a^4}{4c^4}-1\right)^{\frac{1}{2}}\cos 2(\theta+\beta)+\frac{a^2}{2c^2}, \tag{1}$$

Here c and β are arbitrary constants and a is a known constant whose value depends upon the initial speed of the particle. This is the equation of a conic with its centre at O, the centre of force. On writing $a^2 = 2c^2\cosh\alpha$ and eliminating c from (1) we have

$$r^2 = \frac{a^2}{\sinh 2\alpha\cos 2(\theta+\beta)+1+\cosh 2\alpha}, \tag{2}$$

where α and β are now the arbitrary constants. Use (3), § 2.12, in the form appropriate to polar coordinates (as in (3), § 2.16, but with e replaced by α) and take the vectorial angle of A to be zero. Then θ, the vectorial angle of A', the conjugate of A, is given by

$$\frac{\cosh 2\alpha \cos 2(\theta+\beta)+\sinh 2\alpha}{\sin 2(\theta+\beta)} = \frac{\cosh 2\alpha \cos 2\beta+\sinh 2\alpha}{\sin 2\beta}. \tag{3}$$

Simplifying and dividing by $\sin\theta$ we can solve for $\tan\theta$, the result being

$$e^{-4\alpha}\tan(\theta+\beta)\tan\beta = -1. \tag{4}$$

From (4) it can be shown that the tangents to the conic (1) at A and A' must intersect at right angles on the director circle of (2).

The easiest way to prove this is to rotate the initial line of (2) through an angle $-\beta$ about O and then transform (2) to Cartesian coordinates. Its equation becomes

$$e^{2\alpha}x^2+y^2 = \frac{a^2e^\alpha}{2\cosh\alpha}. \tag{5}$$

The angle between the x-axis and OA is now equal to β and that between the x-axis and OA' is equal to $\theta+\beta$. The slopes of the tangents at A and A' are easily found to be $-e^{2\alpha}\cot\beta$ and $-e^{2\alpha}\cot(\theta+\beta)$ respectively. Equation (4) then shows that the two tangents are at right angles to each other and, from known properties of conic sections, it follows that they intersect on the director circle.

The orbit is a circle when $\alpha = 0$ and the equation for θ then reduces to $\theta = \frac{1}{2}\pi$. Thus for the law of direct distance conjugate points on circular orbits subtend a right angle at the centre instead of being diametrically opposite as they are for the inverse square law (see § 2.16).

The final conclusion is that if the angle subtended by AB at the centre of the elliptical orbit is less than θ given by (4) then the action $2\int\limits_A^B T\,dt$ is a weak minimum, otherwise it is stationary but neither maximum nor minimum.

2.18. Properties of solutions of the accessory equation

The accessory equation (1), § 2.4, is a special case of a type of differential equation known as Sturm–Liouville equations.† It may be written in the form

$$\frac{d}{dx}\left(F_{11}\frac{du}{dx}\right)-Gu = 0, \tag{1}$$

where G is an abbreviation for $F_{00}-\frac{d}{dx}(F_{01})$. If u_1 and u_2 are two independent solutions of (1), then

$$u_2\frac{d}{dx}\left(F_{11}\frac{du_1}{dx}\right) = Gu_1 u_2 = u_1\frac{d}{dx}\left(F_{11}\frac{du_2}{dx}\right). \tag{2}$$

From this it follows easily that

$$\frac{d}{dx}\left\{F_{11}\left(u_2\frac{du_1}{dx}-u_1\frac{du_2}{dx}\right)\right\} = 0 \tag{3}$$

and consequently

$$u_2\frac{du_1}{dx}-u_1\frac{du_2}{dx} = \frac{B}{F_{11}}, \tag{4}$$

where B is a constant. We are now in a position to justify an assumption made in the penultimate paragraph of § 2.4, where it was assumed that $u(x)$, a solution of (1), cannot have double roots. For it is evident from (4) that, except in the special case when F_{11} becomes infinite, u_1 and du_1/dx cannot vanish together and therefore that u_1 cannot have a double zero.

From equation (4) it can be deduced further that

$$\frac{d}{dx}\left(\frac{u_1}{u_2}\right) = \frac{B}{F_{11}u_2^2}. \tag{5}$$

Now if the path of integration is not merely an extremal for the integral $\int\limits_A^B F(x,y,y')\,dx$ but makes it a maximum or a minimum, then, as proved in § 2.5, F_{11} must have the same sign at all points of the arc AB. Hence from (5) $d(u_1/u_2)/dx$ has constant sign for all points of arc AB and therefore the expression u_1/u_2 either steadily increases or steadily decreases along this arc. Such a

† E. Lindsay Ince, *Ordinary Differential Equations*, chap. x.

result appears to contradict the requirements of § 2.6, that u_1/u_2 must have the same value at A and at all points conjugate to A. A simple example will show that the contradiction is only apparent. Consider $\tan\theta$, which possesses the following two properties: (i) it steadily increases with θ, because its differential coefficient $\sec^2\theta$ is always positive, and (ii) it satisfies the equality $\tan\theta = \tan(\theta+\pi)$. These two properties can be quite consistent with each other because $\tan\theta$ is not a continuous function for all values of θ. As θ increases from α to $\alpha+\pi$, $\tan\theta$ steadily increases to $+\infty$, and then, as θ passes through an odd multiple of $\frac{1}{2}\pi$, it suffers a sudden drop to $-\infty$, after which it steadily increases again. In general if u_1/u_2 possesses discontinuities it may increase or decrease steadily and yet take the same value for different values of the variable.

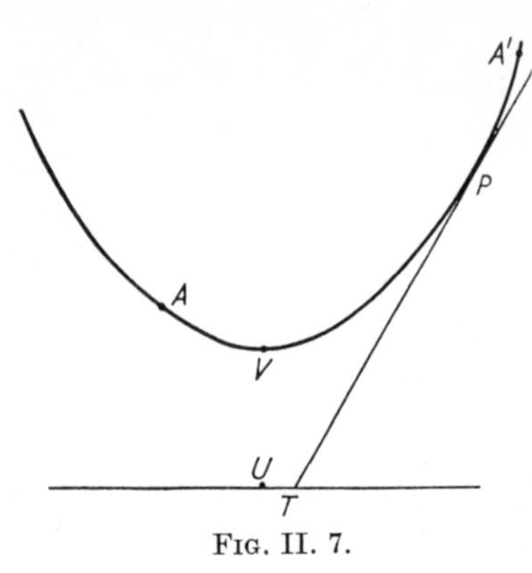

FIG. II. 7.

In the investigations on spherical geodesics, § 2.15, and on orbits under various laws of force, §§ 2.16 and 2.17, the ratio u_1/u_2 is, in each case, expressible as a function of $\tan\theta$. Hence, although the ratio either steadily increases or steadily decreases, it can take the same values at A and at the points conjugate to A.

As a final example, consider the case of the conjugate points on a catenary, § 2.13. If the tangent to the catenary at the point P cuts the directrix at T and if U is the foot of the perpendicular from the vertex V to the directrix, then the ratio u_1/u_2 is proportional to UT, proved in § 2.13. Fig. II. 7 shows the catenary and its directrix, which is also the x-axis, together with the points P, T, U, A and A', the conjugate of A. Consider now the changes in UT as P moves from A to A'. UT steadily increases until P reaches the vertex V, when it becomes infinite. As P moves still farther along the curve T appears on the negative side of the x-axis and it is clear that the algebraical value of UT, taking into account its sign as well as magnitude, steadily increases. It is also clear that as P moves from A to A', T sweeps once across the whole of the x-axis in the positive direction before returning to its original position.

2.19. Summary of the main results of Chapters I and II

Theorem 1. The integral $I = \int_a^b F(x, y, y')\, dx$, whose end points A and B are fixed, is stationary for weak variations, if y satisfies the differential equation (Euler's characteristic equation)

$$\frac{\partial F}{\partial y} - \frac{d}{dx}\left(\frac{\partial F}{\partial y'}\right) = 0. \qquad \S\ 1.4$$

The solutions of Euler's equation are known as extremals.

Variable not explicit in $F(x, y, y')$	*Euler's equation integrates once to*	
x	$F - y'\dfrac{\partial F}{\partial y'} = \text{constant.}$	§ 1.4
y	$\dfrac{\partial F}{\partial y'} = \text{constant.}$	§ 1.4

Legendre's test. If the range of integration (a, b) is sufficiently small and if the sign of $\partial^2 F/\partial y'^2$ is constant throughout this range, then the integral I is a maximum when the sign is negative and a minimum when it is positive. §§ 1.5, 2.5

Characteristic equation identically zero. If the characteristic equation vanishes identically, then I can be evaluated as a function of x and y. § 1.18

Jacobi's test. The accessory equation.

Let $y = s(x)$ be the equation of the extremal which passes through A and B, the end points of the range of integration. Substitute this value of y in $F(x, y, y')$ and let $\partial^2 F/\partial s^2 = F_{00}$, $\partial^2 F/\partial s \partial s' = F_{01}$, and $\partial^2 F/\partial s'^2 = F_{11}$. Then the accessory equation is

$$\left\{F_{00} - \frac{d}{dx}(F_{01})\right\}u - \frac{d}{dx}\left\{F_{11}\frac{du}{dx}\right\} = 0. \qquad \S\ 2.4$$

Conjugate points (kinetic foci): Definition. If (i) $u(x)$ is a solution of the accessory equation, (ii) a is the abscissa of the point A, (iii) $u(a) = 0$, then the roots of $u(x) = 0$ are the abscissae of points on the curve $y = s(x)$ conjugate to A. § 2.6

Theorem 6. If (i) $y = s(x)$ is an extremal through A and B for $I = \int_a^b F(x, y, y')\,dx$, (ii) A' is the first point conjugate to A along the arc AB, (iii) B lies between A and A', (iv) $\partial^2 F/\partial y'^2$ has constant sign for all points of arc AB, then for weak variations I is a maximum when $\partial^2 F/\partial y'^2$ is negative and a minimum when it is positive. § 2.6

Analytical method of finding conjugate points. If (i) a is the abscissa of the point A, (ii) $u_1(x)$ and $u_2(x)$ are independent solutions of the accessory equation, then the equation for the abscissae of the points conjugate to A is

$$\frac{u_1(x)}{u_2(x)} = \frac{u_1(a)}{u_2(a)}. \qquad \S\ 2.6$$

If $y = s(x, c_1, c_2)$ is the general solution of Euler's characteristic equation, then the equation for the abscissae of points conjugate to A is

$$\frac{\partial y}{\partial c_1}\bigg/\frac{\partial y}{\partial c_2} = \left(\frac{\partial y}{\partial c_1}\bigg/\frac{\partial y}{\partial c_2}\right)_{x=a}. \qquad \S\ 2.12$$

These equations can be used with polar and other systems of coordinates if the variables are suitably interpreted.

Geometrical method of finding conjugate points. If the extremals which pass through the point A have an envelope E, then the points of contact of E and the extremal AB are the points conjugate to A. § 2.9

Properties of solutions of the accessory equation. If $u(x)$ is a solution of the accessory equation it cannot have double zeros. § 2.18

If $u_1(x)$ and $u_2(x)$ are independent solutions of this equation, then the ratio $u_1(x)/u_2(x)$ steadily increases or steadily decreases as x increases. § 2.18

CHAPTER III

GENERALIZATIONS OF THE RESULTS OF THE PREVIOUS CHAPTERS

3.1. Introduction

THE results obtained in Chapters I and II can be generalized in several ways, many of which have extremely important applications to problems of pure and applied mathematics. In this chapter three possible methods of generalization will be considered.

The first generalization will deal with integrals of the type

$$I = \int_{t_0}^{t_1} F(q_1, q_2, \ldots, q_n; \dot{q}_1, \dot{q}_2, \ldots, \dot{q}_n; t)\, dt, \tag{1}$$

where $\dot{q}_r$ denotes dq_r/dt. In this integral the variables $q_1, q_2, \ldots, q_n$ are assumed to be independent of each other, in the same way as the parameters which determine the configuration of a body in dynamical problems. For example if $n = 6$, q_1, q_2, q_3 may be the coordinates of G, the centre of gravity of a rigid body, and q_4, q_5, q_6 may be the Eulerian angles† which orientate the body in relation to three mutually perpendicular axes through G. The problem can be restated in the following manner.

Given the functional form of F and choosing the n parameters $q_1, q_2, \ldots, q_n$ to be any arbitrary functions of t, the value of the integral I of equation (1) can then be determined. What forms must these functions of t take in order to render I a maximum or a minimum?

Before the solution is attained the q's are independent of each other, although each q is a dependent variable since it is a function of t, the only independent variable. The independence of the q's is due to the fact that their functional forms are arbitrary until the solution of the problem is attained.

† Ramsey, *Dynamics*, Part II, p. 70.

The second investigation deals with integrals of the type

$$I = \int_a^b F(x, y, y_1, y_2, \dots, y_n)\,dx, \tag{2}$$

where

$$y_r = \frac{d^r y}{dx^r}.$$

Here there is only one independent and only one dependent variable.

The third investigation of this chapter will deal with cases in which there are several independent variables and only one dependent one. An illustration is given by the integral

$$I = \iint F(x, y; z; p, q)\,dxdy. \tag{3}$$

Here the functional form of F and the curves which bound the integral are given, z is a function of x and y, the two independent variables, and $p = \partial z/\partial x$, $q = \partial z/\partial y$.

We shall also deal briefly with integrals of this type in the case when there are three independent variables and one dependent variable, since such integrals occur with great frequency in diverse branches of mathematical physics.

The ideas of the first two chapters are fundamental in each of these investigations. Confining ourselves entirely to the case of weak variations, as in § 1.3, each dependent variable is increased by a quantity of the form $\epsilon\eta(x)$, or by some similar form appropriate to the problem. Here ϵ is a constant and η denotes an arbitrary function of x independent of ϵ. The corresponding variation in I is then evaluated as a power series in ϵ and, as in the elementary theory of maxima and minima, the coefficient of ϵ is equated to zero. In the case of one dependent variable we obtain a characteristic equation, analogous to (8), § 1.4. In the case of an integral with n dependent variables we obtain, in this way, n characteristic equations. On solving the characteristic equation, or equations, we obtain the functional forms of the dependent variables for which I is stationary.

Near a stationary value the variation of I will then depend upon the coefficient of ϵ^2. As in § 2.4, an accessory equation (or equations) can be derived from the characteristic equation (or equations) by means of which this coefficient can be put into a form whose sign can be determined without excessive difficulty. The results, however, are so complex that in this book we shall confine ourselves to the investigation of the first variation of I only, i.e. the coefficient of ϵ. For the second variation, i.e. the coefficient of ϵ^2, we shall either state the relevant results without proof and refer to the advanced literature of the subject or, in very complicated cases, give the references only.

In dynamical problems F is usually a quadratic function of the quantities $\dot{q}_1, \dot{q}_2,..., \dot{q}_n$. This simplification in F is reflected in the second variation which can usually be obtained without excessive difficulty.

3.2. Maxima and minima of integrals of the type

$$I = \int_{t_0}^{t_1} F(q_1, q_2,..., q_n; \dot{q}_1, \dot{q}_2,..., \dot{q}_n; t)\, dt.$$

Since the arguments for two parameters, q_1 and q_2, are fundamentally the same as for n parameters we shall restrict ourselves to the case when I is given by

$$I = \int_{t_0}^{t_1} F(q_1, q_2; \dot{q}_1, \dot{q}_2; t)\, dt. \tag{1}$$

Here the values of t_0 and t_1 and the functional form of F are given. The dot, as usual, denotes differentiation with regard to t.

Let
$$q_1 = s_1(t) = s_1 \tag{2}$$
and
$$q_2 = s_2(t) = s_2 \tag{3}$$
denote the functional forms of the q's which render I stationary. Consider the variation in I due to the q's being varied so that, for a given value of t, q_1 is changed from s_1 to $s_1 + \epsilon_1 u_1$ and q_2 from s_2 to $s_2 + \epsilon_2 u_2$. Here ϵ_1 and ϵ_2 are arbitrary constants independent of q_1, q_2, t and u_1 $(= u_1(t))$ and u_2 $(= u_2(t))$ are arbitrary functions of t, both independent of ϵ_1 and ϵ_2. These are known as weak variations, as in § 1.3. For simplicity we assume that the end points of the range t_0 and t_1 are prescribed, and then $u_1(t_0) = u_1(t_1) = u_2(t_0) = u_2(t_1) = 0$.

Substitute $s_1 + \epsilon_1 u_1$ for q_1 and $s_2 + \epsilon_2 u_2$ for q_2 in (1) and denote the consequent variation in I by δI. Assuming that I possesses continuous partial derivatives with respect to $q_1, q_2, \dot{q}_1$, and $\dot{q}_2$ of at least the second order, we may express the result in powers of ϵ_1 and ϵ_2. The coefficients of ϵ_1 and ϵ_2 in δI are then found to be

$$\int_{t_0}^{t_1} \left\{u_1 \frac{\partial F}{\partial q_1} + \dot{u}_1 \frac{\partial F}{\partial \dot{q}_1}\right\} dt \tag{4}$$

and

$$\int_{t_0}^{t_1} \left\{u_2 \frac{\partial F}{\partial q_2} + \dot{u}_2 \frac{\partial F}{\partial \dot{q}_2}\right\} dt \tag{5}$$

respectively. At a maximum the sign of δI must be negative and at a minimum it must be positive, and in both cases it must be independent of the signs of ϵ_1 and ϵ_2. Hence at a maximum or minimum value of I expressions (4) and (5) must vanish simultaneously.

On integrating the second term of the integrand of (4) by parts and using the conditions $u_1(t_0) = u_2(t_1) = 0$, as in § 1.4, we obtain the equation

$$\int_{t_0}^{t_1} u_1 \left\{\frac{\partial F}{\partial q_1} - \frac{d}{dt}\left(\frac{\partial F}{\partial \dot{q}_1}\right)\right\} dt = 0. \tag{6}$$

Since $u_1(t)$ is an arbitrary function of t, we may apply the arguments of § 1.4 and deduce that (6) can be true only if

$$\frac{\partial F}{\partial q_1} - \frac{d}{dt}\left(\frac{\partial F}{\partial \dot{q}_1}\right) = 0. \tag{7}$$

Similarly from (5) we have

$$\frac{\partial F}{\partial q_2} - \frac{d}{dt}\left(\frac{\partial F}{\partial \dot{q}_2}\right) = 0. \tag{8}$$

These arguments can easily be extended to the case where there are n parameters $q_1, q_2, \ldots, q_n$. The result is:

THEOREM 7. *Let the values of t_0 and t_1 and the functional form of F be given. Then the integral*

$$\int_{t_0}^{t_1} F(q_1, q_2, \ldots, q_n; \dot{q}_1, \dot{q}_2, \ldots, \dot{q}_n; t)\, dt,$$

where the q's are arbitrary functions of t, is stationary for weak variations when the q's satisfy the n equations

$$\frac{\partial F}{\partial q_m}-\frac{d}{dt}\left(\frac{\partial F}{\partial \dot{q}_m}\right) = 0 \quad (m = 1, 2,..., n). \tag{9}$$

3.3. The second variation for integral (1), § 3.2

Reverting to the case when there are only two parameters q_1 and q_2, if equations (7) and (8), § 3.2, are satisfied then the sign of δI will depend upon terms involving ϵ_1^2, ϵ_2^2, and $\epsilon_1 \epsilon_2$. These terms are intricate but can be simplified slightly by assuming that $\epsilon_1 = \epsilon_2$, an assumption which involves no great loss of generality since the functions $u_1(t)$ and $u_2(t)$ still remain arbitrary. With this simplification the second variation of δI depends upon the coefficient of ϵ^2, where $\epsilon_1 = \epsilon_2 = \epsilon$.

A further simplification of the formulae is effected by using the following notation: differentiation with respect to q_1, q_2, $\dot{q}_1$, $\dot{q}_2$ is denoted by the suffixes 1, 2, 3, 4. For example

$$F_{13} = \frac{\partial^2 F}{\partial q_1\, \partial \dot{q}_1}, \qquad F_{14} = \frac{\partial^2 F}{\partial q_1\, \partial \dot{q}_2}, \text{ etc.}$$

On assuming that equations (7) and (8), § 3.2, are satisfied and that I possesses partial derivatives with respect to q_1, q_2, $\dot{q}_1$ and $\dot{q}_2$, of at least the third order, we may use the mean value theorem for functions of two variables. We then obtain†

$$\delta I = \tfrac{1}{2}\epsilon^2 \int\limits_{t_0}^{t_1} \{F_{11} u_1^2 + F_{22} \dot{u}_1^2 + F_{33} u_2^2 + F_{44} \dot{u}_2^2 +$$
$$+2F_{13} u_1 \dot{u}_1 + 2F_{12} u_1 u_2 + 2F_{14} u_1 \dot{u}_2 +$$
$$+2F_{23} \dot{u}_1 u_2 + 2F_{34} \dot{u}_1 \dot{u}_2 + 2F_{24} u_2 \dot{u}_2\}\, dt + O(\epsilon^3). \tag{1}$$

This formula has been studied by several mathematicians, but owing to its complexity we shall state their results without proof.

The ideas of Chapter II can be repeated, although in a more elaborate form. From the two characteristic equations (7) and (8) two accessory equations can be derived, by arguments similar to those of §§ 2.3 and 2.4. With their help (1) can be put into a

† Forsyth, loc. cit., pp. 204–25.

much more tractable form. The final conclusions are that δI has a sign independent of the choice of ϵ and of the arbitrary functions $u_1(t)$, $u_2(t)$ if

(i) the range of integration t_0 to t_1 is sufficiently small,

(ii) throughout the range t_0 to t_1 both F_{22} and F_{44} have constant sign,

(iii) throughout this range $F_{22}F_{44} > F_{24}^2$.

If these conditions are satisfied, then I is a maximum or minimum according as F_{22} is negative or positive (evidently from (iii) F_{22} and F_{44} must both have the same sign).

This result, which is clearly an extension of the Legendre test, §§ 1.5 and 2.5, can be generalized quite easily to the case of n variables, $q_1, q_2,..., q_n$. We then have

THEOREM 8. *Let* $I = \int\limits_{t_0}^{t_1} F(q_1, q_2,..., q_n; \dot{q}_1, \dot{q}_2,..., \dot{q}_n; t)\, dt$, *where the q's satisfy the n second-order partial differential equations* (9), § 3.2. *If at every point of a sufficiently small range t_0 to t_1*

(i) $\dfrac{\partial^2 F}{\partial \dot{q}_r^2}$ $(r = 1, 2,..., n)$ *all have constant sign,*

(ii) $$\left(\frac{\partial^2 F}{\partial \dot{q}_r^2}\right)\left(\frac{\partial^2 F}{\partial \dot{q}_s^2}\right) > \left(\frac{\partial^2 F}{\partial \dot{q}_r \partial \dot{q}_s}\right)^2 \quad \left(\begin{matrix} r = 1, 2,..., n, \\ s = 1, 2,..., n, \end{matrix}\ r \neq s\right),$$

then I is a maximum if the signs of $\partial^2 F/\partial \dot{q}_r^2$ are all negative and minimum if they are all positive (evidently (ii) *ensures that they all have the same sign).*

In dynamics F is a quadratic function of the $\dot{q}$'s and these conditions are frequently fulfilled.

3.4. Conjugate points (kinetic foci) for integral (1), § 3.2

The extent of the range t_0, t_1 for which these results may be true still remains to be determined. The definition and development of conjugate points in §§ 2.6, 2.9, and 2.12 can be generalized so as to be applicable to the case of two or more dependent variables. We give a descriptive account of these generalizations, confining ourselves to the case of two dependent variables since the general formulae are very elaborate.

The theory of differential equations shows that solutions of (7) and (8), § 3.2, are of the form

$$q_1 = \phi_1(t, a_1, a_2, a_3, a_4), \tag{1}$$

$$q_2 = \phi_2(t, a_1, a_2, a_3, a_4), \tag{2}$$

where a_1, a_2, a_3, a_4 are four arbitrary constants independent of t (the same for both (1) and (2)). The elimination of t from these equations leads to a relation between the q's which in a dynamical problem would be the equation of a trajectory or orbit. If two points A and B on such a trajectory are given, then the corresponding values of q_1 and q_2, their coordinates, are known and so the four constants a_1, a_2, a_3, a_4 can be determined. If this determination is unique there will be only one orbit through A and B, but it is also possible that there may be more than one, as in the case of parabolic orbits under gravity. If, however, A only is given, then two of the constants a_1, a_2, a_3, a_4 can be determined and the other two still remain arbitrary. In this case a doubly infinite set of trajectories can be drawn through A, each of which makes the integral I, (1), § 3.2 stationary.

Let T denote a trajectory through A and B and T' any other of the doubly infinite trajectories through A, and let P be a point of intersection of T and T' other than A. Then as T' tends to T the point P will tend to a limiting position, A' say, on T. The point A' is then called the conjugate (or kinetic focus) of A; if there are several such limiting points of intersection the one nearest to A on T will be the one denoted by A' (the illustrations given in §§ 2.10 and 2.11 may be usefully referred to at this stage). The arguments of § 2.6 can then be generalized to show that if B lies between A and A' the solutions of (7) and (8), § 3.2, make the integral I, (1), § 3.2, a maximum or a minimum, and that if B lies outside the arc AA' then, in general, I is stationary only and is neither a maximum nor a minimum.

If the coordinates of A are given, then those of the conjugate point A' can be obtained analytically from an equation which is a generalization of the one given in § 2.12. Let the trajectory be given by equations (1) and (2), let A and its conjugate A'

be points which correspond respectively to the values t_0 and t_0' of t. For brevity let

$$\phi_1(t_0) = \phi_1(t_0, a_1, a_2, a_3, a_4), \quad \phi_2(t_0) = \phi_2(t_0, a_1, a_2, a_3, a_4).$$

Then given t_0 the equation† for t_0' is

$$\begin{vmatrix} \frac{\partial\phi_1(t_0)}{\partial a_1} & \frac{\partial\phi_1(t_0)}{\partial a_2} & \frac{\partial\phi_1(t_0)}{\partial a_3} & \frac{\partial\phi_1(t_0)}{\partial a_4} \\ \frac{\partial\phi_2(t_0)}{\partial a_1} & \frac{\partial\phi_2(t_0)}{\partial a_2} & \frac{\partial\phi_2(t_0)}{\partial a_3} & \frac{\partial\phi_2(t_0)}{\partial a_4} \\ \frac{\partial\phi_1(t_0')}{\partial a_1} & \frac{\partial\phi_1(t_0')}{\partial a_2} & \frac{\partial\phi_1(t_0')}{\partial a_3} & \frac{\partial\phi_1(t_0')}{\partial a_4} \\ \frac{\partial\phi_2(t_0')}{\partial a_1} & \frac{\partial\phi_2(t_0')}{\partial a_2} & \frac{\partial\phi_2(t_0')}{\partial a_3} & \frac{\partial\phi_2(t_0')}{\partial a_4} \end{vmatrix} = 0. \tag{3}$$

The general equation for the conjugate of a point when there are n variables $q_1, q_2,..., q_n$ is given in Forsyth.‡

3.5. Integrals of the type $\int F(x, y, y_1, y_2,..., y_n)\,dx$, where

$$y_m = \frac{d^m y}{dx^m}$$

Since the arguments for the case $n = 2$ can be easily extended to the general value of n, we shall first deal with the integral

$$I = \int_a^b F(x, y, y_1, y_2)\,dx, \tag{1}$$

where the values of a and b and the functional form of F are given. Let

$$y = s(x) \tag{2}$$

be the equation of the curve for which I is stationary. Confining ourselves to weak variations as before, let

$$y = s(x) + \epsilon\eta(x) \tag{3}$$

be the equation of a neighbouring curve, where ϵ is an arbitrary constant and $\eta(x)$ is an arbitrary function of x independent of ϵ.

† Forsyth, *Calculus of Variations*, p. 223.
‡ Loc. cit., pp. 267–70.

We assume that the curves (2) and (3) have common ordinates and tangents at the end points of the range of integration, i.e.

$$\eta(a) = \eta(b) = 0 \text{ and } \left(\frac{d\eta(x)}{dx}\right)_{x=a} = \left(\frac{d\eta(x)}{dx}\right)_{x=b} = 0. \quad (4)$$

We also assume that $F(x, y, y_1, y_2)$ has continuous partial derivatives, of at least the second order, with respect to x, y, y_1, and y_2.

Substituting from (2) and (3) in (1) we obtain for the integral the values I and $I+\delta I$ respectively. On writing

$$s(x) = s, \quad \frac{ds(x)}{dx} = s_1(x), \quad \frac{d^2s(x)}{dx^2} = s_2(x), \quad \eta(x) = \eta,$$

$$\frac{d\eta(x)}{dx} = \eta_1(x), \quad \frac{d^2\eta(x)}{dx^2} = \eta_2(x)$$

and applying the first mean-value theorem we have

$$\delta I = \int_a^b \{F(x, s+\epsilon\eta, s_1+\epsilon\eta_1, s_2+\epsilon\eta_2) - F(x, s, s_1, s_2)\}\, dx$$

$$= \epsilon \int_a^b \left\{\eta \frac{\partial F}{\partial s} + \eta_1 \frac{\partial F}{\partial s_1} + \eta_2 \frac{\partial F}{\partial s_2}\right\} dx + O(\epsilon^2). \quad (5)$$

For a stationary value of I the coefficient of ϵ must vanish and so

$$\int_a^b \left\{\eta \frac{\partial F}{\partial s} + \eta_1 \frac{\partial F}{\partial s_1} + \eta_2 \frac{\partial F}{\partial s_2}\right\} dx = 0. \quad (6)$$

On integrating by parts we have

$$\int_a^b \eta_1 \frac{\partial F}{\partial s_1}\, dx = \eta\left(\frac{\partial F}{\partial s_1}\right)_a^b - \int_a^b \eta \frac{d}{dx}\left(\frac{\partial F}{\partial s_1}\right) dx, \quad (7)$$

and since $\eta(a) = \eta(b) = 0$, the first of the two terms on the right-hand side must vanish at both limits of integration.

Again, on integrating by parts twice we have

$$\int_a^b \eta_2 \frac{\partial F}{\partial s_2}\, dx = \eta_1\left(\frac{\partial F}{\partial s_2}\right)_a^b - \eta \frac{d}{dx}\left(\frac{\partial F}{\partial s_2}\right)_a^b + \int_a^b \eta \frac{d^2}{dx^2}\left(\frac{\partial F}{\partial s_2}\right) dx. \quad (8)$$

Now, by (4), $\eta(a) = \eta(b) = 0$ and $\eta_1(a) = \eta_1(b) = 0$. Hence the first two terms on the right-hand side of (8) vanish at both limits of integration and (6) can be rewritten in the form

$$\int_a^b \eta\left\{\frac{\partial F}{\partial s}-\frac{d}{dx}\left(\frac{\partial F}{\partial s_1}\right)+\frac{d^2}{dx^2}\left(\frac{\partial F}{\partial s_2}\right)\right\}dx = 0. \tag{9}$$

Since $\eta(x)$ is an arbitrary function of x we can apply the methods and arguments of § 1.4 to show that (9) can hold only if

$$\frac{\partial F}{\partial s}-\frac{d}{dx}\left(\frac{\partial F}{\partial s_1}\right)+\frac{d^2}{dx^2}\left(\frac{\partial F}{\partial s_2}\right) = 0. \tag{10}$$

This is a differential equation of the fourth order.

These arguments are easily extended to the case when the integral I contains differential coefficients of the nth order. The result is:

THEOREM 9. *Let $d^m y/dx^m$ be denoted by y_m and let the values of $y, y_1, y_2,..., y_{n-1}$ be given for both $x = a$ and $x = b$. Also let the functional form of F be given; then the integral*

$$I = \int_a^b F(x, y, y_1, y_2,..., y_n)\,dx \tag{11}$$

is stationary when y satisfies the equation

$$\frac{\partial F}{\partial y}-\frac{d}{dx}\left(\frac{\partial F}{\partial y_1}\right)+\frac{d^2}{dx^2}\left(\frac{\partial F}{\partial y_2}\right)-...+(-1)^n\frac{d^n}{dx^n}\left(\frac{\partial F}{\partial y_n}\right) = 0. \tag{12}$$

This equation is a differential equation of order $2n$.

The conditions which must be satisfied in order that these stationary values should be maxima or minima are so elaborate that we cannot deal with them here. A theory of conjugate points can be formulated, analogous to the one developed in Chapter II, and in the case when $n = 2$ it can be proved that if

(i) equation (12) is satisfied when $n = 2$,

(ii) the arc of integration from $x = a$ to $x = b$ does not enclose a point which is conjugate to either of the end points,

(iii) $\partial^2 F/\partial y_2^2$ maintains a constant sign throughout the arc of integration from $x = a$ to $x = b$,

then I is a maximum if this sign is negative and a minimum if it is positive.

Proofs of these statements and formulae for conjugate points can be found in Forsyth's *Calculus of Variations*.†

3.6. The case of several independent variables and one dependent variable

We now deal with integrals of the type

$$I = \iint F(x, y; z; p; q)\,dxdy, \tag{1}$$

where x and y are independent variables, z is a function of x and y, and $p = \partial z/\partial x$ and $q = \partial z/\partial y$. The functional form of F is given and the problem consists in finding the functional relationship between x, y, and z which renders I a maximum or minimum. Such a relationship can be expressed graphically by a surface in three-dimensional space. Surfaces for which I is stationary will be referred to as extremals. In order to simplify the analysis we shall assume that the curves which bound the area of integration are fixed.

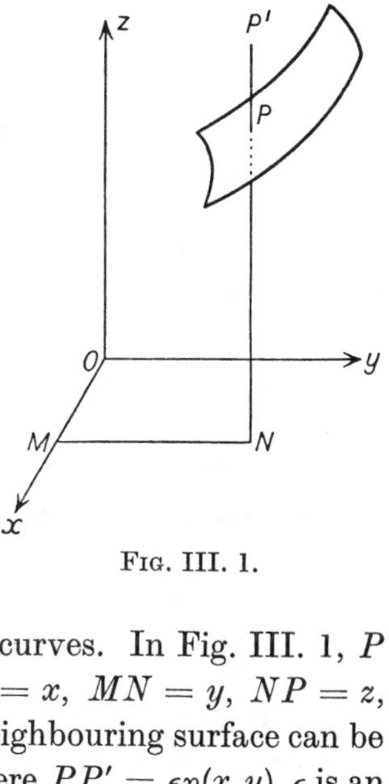

Let $$z = s(x, y) \tag{2}$$

Fig. III. 1.

be the equation of an extremal surface passing through the given boundary curves. In Fig. III. 1, P is a point on this surface and $OM = x$, $MN = y$, $NP = z$, where (x, y, z) are its coordinates. A neighbouring surface can be obtained by producing NP to P', where $PP' = \epsilon\eta(x, y)$, ϵ is an arbitrary constant, and $\eta(x, y)$ is an arbitrary function of x and y. We confine ourselves to weak variations, as in § 1.3, and for this purpose we assume that $\eta(x, y)$ is independent of ϵ. The equation of the neighbouring surface is

$$z = s(x, y) + \epsilon\eta(x, y). \tag{3}$$

Since the boundaries of integration of I are fixed, then at all points of the bounding curves we must have $\eta(x, y) = 0$.

† Chap. iii.

Finally we assume that $F(x,y,z,p,q)$ possesses continuous partial derivatives with respect to z, p, and q; this enables us to utilize a mean-value theorem.†

For brevity write $s(x,y) = s$, $\eta(x,y) = \eta$, and let I denote the value of integral (1) when taken over the extremal (2) and $I+\delta I$ its value when taken over the surface (3). Then we have

$$\delta I = \iint F\left(x, y, s+\epsilon\eta, p+\epsilon\frac{\partial\eta}{\partial x}, q+\epsilon\frac{\partial\eta}{\partial y}\right) dxdy - \\ - \iint F(x,y,s,p,q)\, dxdy$$

$$= \epsilon \iint \left(\eta\frac{\partial F}{\partial s}+\frac{\partial\eta}{\partial x}\frac{\partial F}{\partial p}+\frac{\partial\eta}{\partial y}\frac{\partial F}{\partial q}\right) dxdy + O(\epsilon^2), \quad (4)$$

where p and q denote values of $\partial z/\partial x$ and $\partial z/\partial y$ on the surface (2). For a maximum or a minimum the coefficient of ϵ must vanish and so

$$\iint \left(\eta\frac{\partial F}{\partial s}+\frac{\partial\eta}{\partial x}\frac{\partial F}{\partial p}+\frac{\partial\eta}{\partial y}\frac{\partial F}{\partial q}\right) dxdy = 0. \quad (5)$$

By methods similar to those employed in § 1.4 we can deduce from (5) a second-order partial differential equation, but before doing so we must first generalize the principle of integration by parts to a form suitable for application to the process of double integration.

For this purpose we prove the following lemma on double integration. This is usually known as Cauchy's integral theorem.

3.7. Lemma on double integration

Fig. III.2 shows the closed curve $ABCD$ in the (x,y) plane, a curve denoted by Γ. Let $\phi(x,y)$ $(\equiv \phi)$ and $\psi(x,y)$ $(\equiv \psi)$ be two functions whose first-order partial derivatives with respect to x and y are continuous functions of both these variables. Then

$$\iint \left(\frac{\partial\phi}{\partial x}+\frac{\partial\psi}{\partial y}\right) dxdy = \int (\phi\, dy - \psi\, dx), \quad (1)$$

where (i) the double integral is taken over the area bounded by Γ, (ii) the line integral is taken along the curve Γ, and (iii) the direction of integration round Γ keeps the area of integration on the left-hand side.

† G. A. Gibson, *Advanced Calculus*, p. 103 (Macmillan).

Proof. If AC, Fig. 3.2, is parallel to the x-axis then, on integrating with respect to x, we have

$$\iint \frac{\partial \phi}{\partial x}\,dxdy = \int (\phi_c\,dy_c - \phi_a\,dy_a), \tag{2}$$

where the subscript a denotes that the variables in $\phi_a\,dy_a$ take values corresponding to the point A. A corresponding interpretation applies to the subscript c.

Now on traversing Γ so that the area is kept on the left-hand side we have $dy_c = +dy$ and $dy_a = -dy$ (Fig. III. 2). Also it is evident that as the chord AC varies from the position in which its distance from the x-axis is a minimum to one in which its distance is a maximum, the points A and C describe different arcs of Γ in such a manner that between them Γ is described once. Hence

$$\iint \frac{\partial \phi}{\partial x}\,dxdy = \int (\phi_c\,dy_c - \phi_a\,dy_a) = \int_\Gamma \phi\,dy, \tag{3}$$

where the last integral is taken once round Γ.

Again, if BD is parallel to the y-axis (Fig. III. 2), then, on integrating with respect to y, we have

$$\iint \frac{\partial \psi}{\partial y}\,dxdy = \int (\psi_d\,dx_d - \psi_b\,dx_b).$$

But if Γ is described so as to keep the area on the left-hand side, then $dx_d = -dx$ and $dx_b = +dx$. Thus, similarly to (3), we have

$$\iint \frac{\partial \psi}{\partial y}\,dxdy = -\int_\Gamma \psi\,dx. \tag{4}$$

On adding (3) and (4) we obtain (1).

Fig. III. 2 illustrates a simply connected region, i.e. one in which all curves of the region can be shrunk into points without having to cross the bounding-curve Γ. The lemma is still true for multiply connected regions, of which an illustration is given by Fig. III. 3. Here the boundary region consists of two closed ovals one lying entirely inside the other, $ABCD$ and $A'B'C'D'$. The curve Γ consists of the two ovals taken together and, in order to satisfy the convention that Γ is to be described so that the area is kept on the left-hand side, each of these ovals must be described in the sense indicated by the arrows in the diagram. On integrating with respect to x and taking $AA'C'C$ in Fig. III. 3 to be parallel to the x-axis, we have

$$\iint \frac{\partial \phi}{\partial x}\, dxdy = \int (\phi_{a'}\, dy_{a'} - \phi_a\, dy_a + \phi_c\, dy_c - \phi_{c'}\, dy_{c'}). \qquad (5)$$

Here $dy_{a'} = dy_c = +dy$ and $dy_a = dy_{c'} = -dy$. Evidently as $AA'C'C$ varies from its minimum to its maximum distance

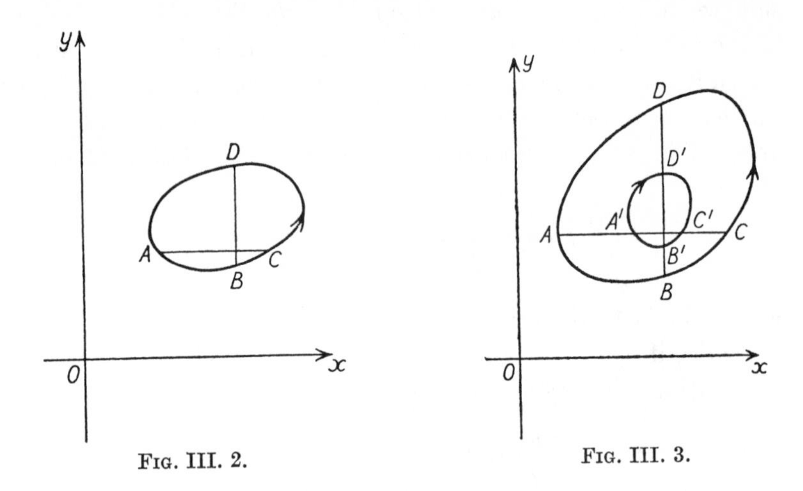

FIG. III. 2. FIG. III. 3.

from the x-axis it is clear that the four points $AA'C'C$ traverse four different arcs of the boundary and that these four arcs add up to Γ once, without overlap. Hence (5) can be written

$$\iint \frac{\partial \phi}{\partial x}\, dxdy = \int_{\Gamma} \phi\, dy, \qquad (6)$$

where the right-hand integral is taken once round Γ. Similar arguments applied to the integral $\iint (\partial\psi/\partial y)\, dxdy$ then enable us to complete the proof of the lemma for the case of multiply connected regions. This lemma is a special case of Stokes's theorem and further details are available in most works on analysis.†

† Courant, *Differential and Integral Calculus*, vol. ii, chap. v; Rutherford, *Vector Methods*, p. 75.

3.8. The characteristic equation for the integral (1), § 3.6

On writing $\phi = \eta \dfrac{\partial F}{\partial p}$ and $\psi = \eta \dfrac{\partial F}{\partial q}$ in (1), § 3.7, we have

$$\iint \left[\eta\left\{\frac{\partial}{\partial x}\left(\frac{\partial F}{\partial p}\right)+\frac{\partial}{\partial y}\left(\frac{\partial F}{\partial q}\right)\right\}+\frac{\partial \eta}{\partial x}\frac{\partial F}{\partial p}+\frac{\partial \eta}{\partial y}\frac{\partial F}{\partial q}\right] dxdy$$
$$= \int \eta\left(\frac{\partial F}{\partial p}\,dy-\frac{\partial F}{\partial q}\,dx\right), \quad (1)$$

where the line integral is taken round the given boundaries of the integral (1), § 3.6. Since these boundaries are the same for surfaces (2) and (3), § 3.6, we must have $\eta = 0$ in the line integral on the right-hand side of (1). Hence (1) reduces to

$$\iint \left(\frac{\partial \eta}{\partial x}\frac{\partial F}{\partial p}+\frac{\partial \eta}{\partial y}\frac{\partial F}{\partial q}\right) dxdy = -\iint \eta\left\{\frac{\partial}{\partial x}\left(\frac{\partial F}{\partial p}\right)+\frac{\partial}{\partial y}\left(\frac{\partial F}{\partial q}\right)\right\} dxdy, \quad (2)$$

and so equation (5), § 3.6, reduces to

$$\iint \eta\left\{\frac{\partial F}{\partial s}-\frac{\partial}{\partial x}\left(\frac{\partial F}{\partial p}\right)-\frac{\partial}{\partial y}\left(\frac{\partial F}{\partial q}\right)\right\} dxdy = 0. \quad (3)$$

We can now make use of the fact that over the surface of integration $\eta = \eta(x,y)$ is an arbitrary function of x and y and apply the arguments of § 1.4. We then deduce that (3) can be true only if

$$\frac{\partial F}{\partial s}-\frac{\partial}{\partial x}\left(\frac{\partial F}{\partial p}\right)-\frac{\partial}{\partial y}\left(\frac{\partial F}{\partial q}\right) = 0. \quad (4)$$

This is the characteristic equation for the determination of the maximum and minimum values of the integral (1), § 3.6. It is a partial differential equation of the second order for the determination of z $(= s(x,y))$ as a function of x and y.

The ideas of these three paragraphs, including the lemma of § 3.7, are easily extended to the case of n independent variables $x_1, x_2,..., x_n$ and one dependent variable z. We then deduce

THEOREM 10. *Let F be a given functional form, let $p_r = \partial z/\partial x_r$ where z is a function of $x_1, x_2,..., x_n$, and let the integral*

$$I = \iint ... \int F(z, x_1, x_2,..., x_n, p_1, p_2,..., p_n)\, dx_1 dx_2... dx_n \quad (5)$$

be taken through an n-dimensional region bounded by given fixed *boundaries of dimension* $n-1$. *Then the integral* (5) *is stationary when* z *is a solution of the second-order partial differential equation*

$$\frac{\partial F}{\partial z}-\frac{\partial}{\partial x_1}\left(\frac{\partial F}{\partial p_1}\right)-\frac{\partial}{\partial x_2}\left(\frac{\partial F}{\partial p_2}\right)-\dots-\frac{\partial}{\partial x_n}\left(\frac{\partial F}{\partial p_n}\right)=0. \tag{6}$$

3.9. The second variation of integral (1), § 3.6

The study of the second variation of the integral in (1), § 3.6 involves elaborate analysis. We limit ourselves to a summary, without proof, of the results obtained for the case of two and also of three independent variables. For proofs the reader is referred to Forsyth.†

From the characteristic equation (4), § 3.8, we can deduce an accessory equation by means of which a theory of conjugate curves or surfaces can be developed, analogous to the theory of conjugate points in § 2.6. Suppose that the domain of integration of the integral (1), § 3.6, on an extremal surface is bounded by two closed non-intersecting curves Γ and Δ. Then it is possible to find curves conjugate to Γ and Δ either by analytical or by geometrical methods. For example, if the family of extremals passing through Γ possesses an envelope E, then the curve of contact of E and the extremal through both Γ and Δ is conjugate to Γ. In the case of two independent variables it can be proved that if

(i) the domain of integration on the extremal surface does not contain points on the conjugate of either of its bounding curves,

(ii) at all points of this domain,

(ii *a*) $\partial^2F/\partial p^2$ has constant sign and

(ii *b*) $\dfrac{\partial^2F}{\partial p^2}\dfrac{\partial^2F}{\partial q^2}>\left(\dfrac{\partial^2F}{\partial p\partial q}\right)^2$,

then the solution of

$$\frac{\partial F}{\partial z}-\frac{\partial}{\partial x}\left(\frac{\partial F}{\partial p}\right)-\frac{\partial}{\partial y}\left(\frac{\partial F}{\partial q}\right)=0, \tag{1}$$

considered as an equation for z, makes $\iint F(x,y,z,p,q)\,dxdy$ a

† Loc. cit., chap. ix.

maximum if $\partial^2 F/\partial p^2$ is negative and a minimum if it is positive. From (ii *b*) $\partial^2 F/\partial p^2$ and $\partial^2 F/\partial q^2$ have the same sign.

In the case of three independent variables the integral to be considered is

$$I = \iiint F(x_1, x_2, x_3, z, p_1, p_2, p_3)\, dxdydz, \tag{2}$$

where x_1, x_2, x_3 are the independent variables and $p_1 = \partial z/\partial x_1$, $p_2 = \partial z/\partial x_2$, $p_3 = \partial z/\partial x_3$. The result just stated for the case of two independent variables remains true for the integral (2) if (1) is replaced by (6), § 3.8 with $n = 3$. Condition (i) still remains and (ii) is replaced by

(ii *c*) throughout the domain of integration $\partial^2 F/\partial p_1^2$ has constant sign, and

(ii *d*) the following three inequalities hold:

$$\frac{\partial^2 F}{\partial p_2^2}\frac{\partial^2 F}{\partial p_3^2} > \left(\frac{\partial^2 F}{\partial p_2\,\partial p_3}\right)^2, \qquad \frac{\partial^2 F}{\partial p_3^2}\frac{\partial^2 F}{\partial p_1^2} > \left(\frac{\partial^2 F}{\partial p_3\,\partial p_1}\right)^2,$$

$$\frac{\partial^2 F}{\partial p_1^2}\frac{\partial^2 F}{\partial p_2^2} > \left(\frac{\partial^2 F}{\partial p_1\,\partial p_2}\right)^2,$$

then the integral (2) is a maximum if $\partial^2 F/\partial p_1^2$ is negative and a minimum if it is positive. From (ii *d*) it is evident that $\partial^2 F/\partial p_1^2$, $\partial^2 F/\partial p_2^2$, $\partial^2 F/\partial p_3^2$ all have the same sign.

3.10. Applications to physical and other problems

The most important application of these results is to integrals of the type

$$I = \frac{1}{8\pi}\iiint \left\{\left(\frac{\partial v}{\partial x}\right)^2 + \left(\frac{\partial v}{\partial y}\right)^2 + \left(\frac{\partial v}{\partial z}\right)^2\right\} dxdydz, \tag{1}$$

where x, y, z are independent variables and v is the dependent variable. On writing $x = x_1$, $y = x_2$, $z = x_3$, $\partial v/\partial x = p_1$, $\partial v/\partial y = p_2$, and $\partial v/\partial z = p_3$ so as to make use of the terminology of § 3.8, we have

$$F(z, x_1, x_2, x_3, p_1, p_2, p_3) = \frac{1}{8\pi}(p_1^2+p_2^2+p_3^2). \tag{2}$$

Equation (6), § 3.8 then becomes

$$-\frac{1}{4\pi}\left\{\frac{\partial p_1}{\partial x}+\frac{\partial p_2}{\partial y}+\frac{\partial p_3}{\partial z}\right\} = 0, \tag{3}$$

i.e.

$$\frac{\partial^2 v}{\partial x^2}+\frac{\partial^2 v}{\partial y^2}+\frac{\partial^2 v}{\partial z^2} = 0. \tag{4}$$

Also $$\frac{\partial^2 F}{\partial p_1^2} = \frac{\partial^2 F}{\partial p_2^2} = \frac{\partial^2 F}{\partial p_3^2} = \frac{1}{4\pi}$$

and $$\frac{\partial^2 F}{\partial p_2 \partial p_3} = \frac{\partial^2 F}{\partial p_3 \partial p_1} = \frac{\partial^2 F}{\partial p_1 \partial p_2} = 0.$$

Hence from (ii c) and (ii d), § 3.9, the integral (1) admits minimum values for solutions of (4).

For physical interpretations let v be the potential function arising from a distribution of matter attracting according to the Newtonian law of gravity† or from a distribution of electric charge in electrostatic equilibrium‡ or from a magnetic distribution,§ then integral (1) gives the energy density in space and (4) is Laplace's equation. In the case of an electrostatic distribution Laplace's equation holds throughout space, showing that the charge distributes itself in such a manner that the energy is a minimum.

In hydrodynamics if v is the velocity potential of a fluid in irrotational motion and if the fluid density is independent of the time (i.e. if the motion is steady), then the integral (1) is proportional to the kinetic energy of the system and (4) is the equation of continuity, which holds throughout space. Thus the motion adjusts itself so that the kinetic energy is a minimum.‖

3.11. Application to theory of minimal surfaces

A minimal surface is one whose area, when bounded by two given closed non-intersecting curves, is a minimum. The area of a surface whose equation is

$$z = f(x, y) \tag{1}$$

is given by†† $$\iint (1+p^2+q^2)^{\frac{1}{2}}\, dxdy, \tag{2}$$

where $p = \partial z/\partial x$ and $q = \partial z/\partial y$.

To minimize this integral we have

$$F(z, x, y, p, q) = (1+p^2+q^2)^{\frac{1}{2}}, \tag{3}$$

† Ramsey, *Theory of Newtonian Attraction*, §§ 5.3 and 4.1.

‡ Jeans, *The Mathematical Theory of Electricity and Magnetism*, § 168.

§ Jeans, loc. cit., § 451.

‖ Ramsey, *A Treatise on Hydromechanics*, vol. ii, §§ 4.6 and 4.7.

†† Courant, *Differential and Integral Calculus*, vol. ii, pp. 268 et. seq.

so that the characteristic equation, (6), § 3.8, is

$$-\frac{\partial}{\partial x}\left\{\frac{p}{(1+p^2+q^2)^{\frac{1}{2}}}\right\}-\frac{\partial}{\partial y}\left\{\frac{q}{(1+p^2+q^2)^{\frac{1}{2}}}\right\} = 0. \tag{4}$$

On writing $1+p^2+q^2 = E^2$, differentiating and multiplying through by E^3, we obtain

$$\frac{\partial p}{\partial x}E^2-p\left(p\frac{\partial p}{\partial x}+q\frac{\partial q}{\partial x}\right)+\frac{\partial q}{\partial y}E^2-q\left(p\frac{\partial p}{\partial y}+q\frac{\partial q}{\partial y}\right) = 0. \tag{5}$$

This easily simplifies to

$$(1+q^2)\frac{\partial^2 z}{\partial x^2}-2pq\frac{\partial^2 z}{\partial x\partial y}+(1+p^2)\frac{\partial^2 z}{\partial y^2} = 0. \tag{6}$$

Equation (6), which is the well-known differential equation of minimal surfaces, was first obtained by Lagrange in 1760 and has been much studied since. It expresses in analytical form the geometrical property that the sum of the two principal curvatures at any point of a minimal surface is zero.† This property is the one generally used to define a minimal surface, for it has the advantage that it is independent of any boundary conditions.

A simple illustration of a minimal surface is given by the rotation of a catenary about its directrix. Taking the z-axis as the directrix and a suitable scale of measurement, the equation of the resulting surface of revolution is

$$(x^2+y^2)^{\frac{1}{2}} = \cosh z, \tag{7}$$

where the positive value of the root is taken.

Squaring (7), differentiating partially with respect to x and y, and writing $p = \partial z/\partial x$, $q = \partial z/\partial y$ we have

$$x = p\cosh z\sinh z, \qquad y = q\cosh z\sinh z \tag{8}$$

$$\left.\begin{aligned} 1 &= \frac{\partial^2 z}{\partial x^2}\cosh z\sinh z+p^2\cosh 2z \\ 0 &= \frac{\partial^2 z}{\partial x\partial y}\cosh z\sinh z+pq\cosh 2z \\ 1 &= \frac{\partial^2 z}{\partial y^2}\cosh z\sinh z+q^2\cosh 2z \end{aligned}\right\}. \tag{9}$$

† R. J. T. Bell, *Coordinate Geometry of Three Dimensions*, p. 337, § 232.

Denoting the left-hand side of (6) by L we have from (9)

$$L\cosh z\sinh z = (1+q^2)(1-p^2\cosh 2z)+ \\ +2p^2q^2\cosh 2z+(1+p^2)(1-q^2\cosh 2z) \quad (10)$$

$$= 2-2(p^2+q^2)\sinh^2 z. \quad (11)$$

Using (8) and then (7) we have finally

$$L\cosh^3 z\sinh z = 2\cosh^2 z-2(x^2+y^2) = 0. \quad (12)$$

Hence the surface whose equation is (7) satisfies (6) and is a minimal surface.

In order to apply the tests of (ii), § 3.9, we find that

$$\frac{\partial^2 F}{\partial p^2} = \frac{1+q^2}{(1+p^2+q^2)^{\frac{3}{2}}}; \quad \frac{\partial^2 F}{\partial p\partial q} = -\frac{pq}{(1+p^2+q^2)^{\frac{3}{2}}};$$

$$\frac{\partial^2 F}{\partial q^2} = \frac{1+p^2}{(1+p^2+q^2)^{\frac{3}{2}}}.$$

It is then evident that

$$\frac{\partial^2 F}{\partial p^2}\frac{\partial^2 F}{\partial q^2} > \left(\frac{\partial^2 F}{\partial p\partial q}\right)^2.$$

Since the positive value of the root is taken it follows that $\partial^2 F/\partial p^2$ is always positive and so the surface admits a minimum.

The problem of finding the conjugate of a given curve is naturally a difficult one, but the following result, which we quote without proof, has been obtained in a case with simplified conditions. Call the intersection of surface (7) with the plane $z = h$, where h is constant, a circle of latitude. Then the conjugate of a circle of latitude on the plane $z = h_1$ is another circle of latitude which lies on the plane $z = h_2$, where†

$$\coth h_1 - h_1 = \coth h_2 - h_2. \quad (13)$$

This equation is the same as (3), § 2.13, and the same geometrical consequences must therefore follow.

Consider the problem of constructing a minimal surface, such as (7), to pass through two given circles C_1 and C_2 lying in planes perpendicular to O_1O_2, the join of their centres, Fig. III. 4.

† Forsyth, loc. cit., p. 480.

If a plane through $O_1 O_2$ cuts the circles in the points P_1 and P_2, then at most two catenaries can be drawn to pass through P_1 and P_2 and to have $O_1 O_2$ as directrix.† On revolving these catenaries round $O_1 O_2$ as axis it follows that at most two surfaces, such as (7), can be constructed to pass through C_1 and C_2. Denote one of these surfaces by S and by means of (13) find C_1' and C_2', the respective conjugates of C_1 and C_2. If the region of S bounded by C_1 and C_2 does not contain either C_1' or C_2', then for any surface of revolution bounded by C_1 and C_2 the area of S is a minimum.

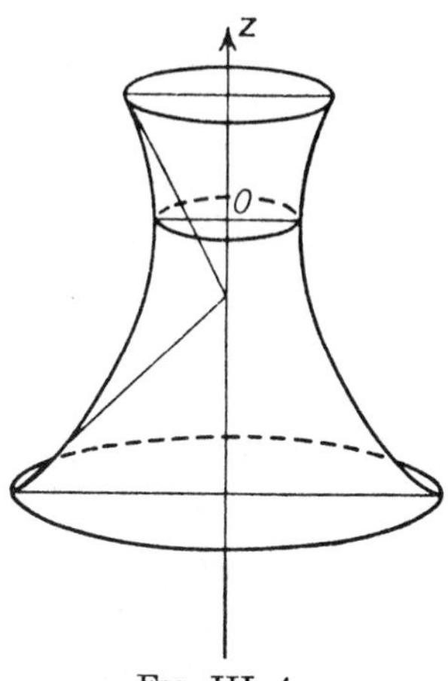

FIG. III. 4.

In this section the theory of minimal surfaces has been considered in relation to three-dimensional space, but the idea can be extended to a hyperspace of any number of dimensions. The case of four dimensions has been dealt with by Forsyth.‡

EXAMPLE. Find the minimal surface whose equation is expressible in the form $z = f(x)+F(y)$.

(Answer $e^{az} = \cos ax \sec ay$, where a is any constant.)

† Forsyth, loc. cit., p. 100.
‡ Forsyth, loc. cit., p. 643.

CHAPTER IV

RELATIVE MAXIMA AND MINIMA AND ISOPERIMETRICAL PROBLEMS

4.1. Introduction

THE problems dealt with in this chapter are illustrated by the following two examples. (i) Given the length of a closed plane curve, find its shape when the enclosed area is a maximum. Expressed otherwise we require the maximum value of $\int dA$ subject to the condition $\int ds = L$, where L is the given length, ds is the element of arc, and dA the element of area. (ii) Find the curve, lying wholly on a given surface S, whose length of arc between two given points P and Q is a minimum. Here we require the minimum value of $\int ds$ subject to the condition that $S(x, y, z) = 0$, where ds is the element of arc and the equation is that of the surface S.

Problems in which the conditional equation involves integration are usually known as isoperimetrical problems.

4.2. Relative maxima and minima

Confining ourselves to plane curves we commence with the problem of finding the stationary values of the integral I, where

$$I = \int_a^b F(x, y, y_1)\, dx, \tag{1}$$

and where the equation

$$L = \int_a^b \phi(x, y, y_1)\, dx \tag{2}$$

must be satisfied by y and y_1 $(= dy/dx)$ regarded as functions of x. Here L is a given constant and F and ϕ are given functional forms. We may express the problem otherwise by saying that among all the curves which satisfy (2) we require those which maximize or minimize (1). These maximizing (or minimizing) curves will be referred to as extremals.

For simplicity we shall consider only the case where the end points of the range of integration, $x = a$ and $x = b$, are prescribed. We also assume that F and ϕ possess continuous partial derivatives up to at least the second order, so that we may employ the mean-value theorem for functions of several variables.

With the terminology of § 1.3 let

$$y = s(x) \tag{3}$$

be the functional form of y for which I is stationary and let us consider the change in the value of I when y is subjected to a variation of the form

$$y = s(x)+\epsilon t(x). \tag{4}$$

Here ϵ is a constant and, since the end points are prescribed, $t(a) = t(b) = 0$. If I changes to $I+\delta I$ then, using the mean-value theorem, we have

$$\delta I = \epsilon \int\limits_a^b \left\{\frac{\partial F}{\partial s} t(x)+\frac{\partial F}{\partial s_1} t_1(x)\right\} dx+O(\epsilon^2), \tag{5}$$

where $s = s(x)$, $s_1 = ds(x)/dx$, $t_1(x) = dt(x)/dx$ and, using the Landau notation of § 1.2, $O(\epsilon^2)$ denotes terms containing ϵ^2 and higher powers of ϵ.

Since L is constant, on dealing with integral (2) similarly we have

$$0 = \epsilon \int\limits_a^b \left\{\frac{\partial \phi}{\partial s} t(x)+\frac{\partial \phi}{\partial s_1} t_1(x)\right\} dx+O(\epsilon^2). \tag{6}$$

Now integrate by parts the second term in each of these integrals and make use of the conditions $t(a) = t(b) = 0$. Equations (5) and (6) then respectively become

$$\delta I = \epsilon \int\limits_a^b t(x)\left\{\frac{\partial F}{\partial s}-\frac{d}{dx}\left(\frac{\partial F}{\partial s_1}\right)\right\} dx+O(\epsilon^2) \tag{7}$$

and

$$0 = \epsilon \int\limits_a^b t(x)\left\{\frac{\partial \phi}{\partial s}-\frac{d}{dx}\left(\frac{\partial \phi}{\partial s_1}\right)\right\} dx+O(\epsilon^2). \tag{8}$$

From (8) it is evident that ϵ and $t(x)$ are not independent of each other and therefore we may not equate the coefficient of ϵ in (7) to zero, as in previous work. A similar difficulty occurs when finding the stationary values of functions of several variables in the case where some of the variables satisfy subsidiary conditions. Among the methods used to overcome this difficulty is the Lagrangian method of undetermined multipliers† which, fortunately, applies with equal success to the problems of this chapter.

Multiplying equation (8) by an undetermined multiplier λ and subtracting from (7) we have

$$\delta I = \epsilon\left[\int\limits_a^b t(x)\left\{\frac{\partial F}{\partial s}-\frac{d}{dx}\left(\frac{\partial F}{\partial s_1}\right)\right\}dx-\lambda\int\limits_a^b t(x)\left\{\frac{\partial \phi}{\partial s}-\frac{d}{dx}\left(\frac{\partial \phi}{\partial s_1}\right)\right\}dx\right]+ +O(\epsilon^2). \tag{9}$$

Now if I is to be stationary the sign of δI must be independent of the choice of ϵ and $t(x)$. This result can be achieved if we assume that λ is a constant and choose $s(x)$ so as to satisfy the second-order differential equation

$$\frac{\partial}{\partial s}(F-\lambda\phi)-\frac{d}{dx}\left\{\frac{\partial}{\partial s_1}(F-\lambda\phi)\right\} = 0. \tag{10}$$

It might appear sufficient to choose $s(x)$ so as to satisfy the second-order equation

$$\frac{\partial F}{\partial s}-\frac{d}{dx}\left(\frac{\partial F}{\partial s_1}\right) = 0. \tag{11}$$

For it would then follow from (7) that the sign of δI is independent of ϵ and $t(x)$. But only a finite number of solutions of (11) can pass through the fixed end points A and B, and none of these may satisfy condition (2). In the case of equation (10), λ can be regarded as a third arbitrary constant in addition to the two of the solution; hence a solution of (10) can be found which passes through the end points A and B and in addition satisfies condition (2).

Evidently the conditions under which (10) has been proved are sufficient but not necessary.

† Courant, *Differential and Integral Calculus*, vol. ii, p. 188 et seq.

To summarize our results we have proved:

THEOREM 11. *By suitable choice of the constant* λ, *a solution of the second-order equation*

$$\frac{\partial}{\partial y}(F-\lambda\phi)-\frac{d}{dx}\left\{\frac{\partial}{\partial y_1}(F-\lambda\phi)\right\}=0 \tag{12}$$

can be found which renders the integral I, *of equation* (1), *stationary, which passes through the end points of the range of integration of* (1) *and* (2), *and which also satisfies condition* (2).

The discussion of the second variation, i.e. the coefficient of ϵ^2, is postponed to § 4.12; the next three sections are devoted to illustrations of the theorem.

4.3. Examples illustrating theorem 11

EXAMPLE 1. Two fixed points, A and B, are joined by a plane curve Γ, of given length l. Find the form of Γ for which the area enclosed by Γ and the chord AB is a maximum.

It is best to use polar coordinates, taking AB as the initial line and some convenient point O, lying between A and B, as pole. It is assumed that a radius vector through O cuts Γ in at most one point.

The problem evidently becomes that of maximizing the integral

$$I=\tfrac{1}{2}\int\limits_0^\pi r^2\,d\theta, \tag{1}$$

subject to the condition that

$$l=\int\limits_A^B ds \tag{2}$$

or

$$l=\int\limits_0^\pi\left\{r^2+\left(\frac{dr}{d\theta}\right)^2\right\}^{\frac{1}{2}}d\theta, \tag{3}$$

where the positive value of the root is to be taken.

Here θ is taken as the independent variable and r as the dependent variable (instead of x and y respectively). On writing $r_1=dr/d\theta$ we have $F(\theta,r,r_1)=\frac{1}{2}r^2$ and $\phi(\theta,r,r_1)=(r^2+r_1^2)^{\frac{1}{2}}$. Hence with x replaced by θ and y by r, (12), § 4.2 becomes

$$r-\frac{\lambda r}{(r^2+r_1^2)^{\frac{1}{2}}}+\frac{d}{d\theta}\left\{\frac{\lambda r_1}{(r^2+r_1^2)^{\frac{1}{2}}}\right\}=0. \tag{4}$$

If s denotes the length of arc we have $(r^2+r_1^2)^{\frac{1}{2}} = ds/d\theta$, and so, since λ is constant, equation (4) can be written

$$r-\lambda r\frac{d\theta}{ds}+\lambda\frac{d}{d\theta}\left(\frac{dr}{ds}\right) = 0. \tag{5}$$

If χ is the angle between the radius vector OP and the tangent to Γ at P (Fig. IV. 1) and ψ is the angle between the initial line and this tangent we have $r\dfrac{d\theta}{ds} = \sin\chi$ and $\dfrac{dr}{ds} = \cos\chi$.

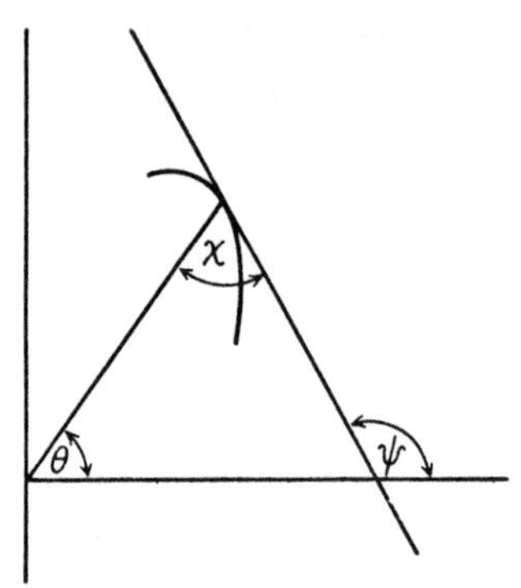

FIG. IV. 1.

Equation (5) can be written

$$r = \lambda\left\{\sin\chi-\frac{d}{d\theta}(\cos\chi)\right\} = \lambda\sin\chi\left\{1+\frac{d\chi}{d\theta}\right\}$$

$$= \lambda r\frac{d\theta}{ds}\frac{d\psi}{d\theta} = \lambda r\frac{d\psi}{ds}. \tag{6}$$

Thus $\lambda = ds/d\psi$ and, since λ is constant, it follows that at all points of Γ the radius of curvature must be the same. Hence in the stationary case Γ is an arc of a circle of radius λ.

If $AB = 2a$ and the circular arc AB subtends an angle 2α at its centre C, it is easy to show geometrically that $a = \lambda\sin\alpha$ and $l = 2\alpha\lambda$. Eliminating λ we have

$$2a\alpha = l\sin\alpha, \tag{7}$$

giving us an equation for α in terms of the known quantities l and a. Equation (7) always has a real solution if $l > 2a$, and if this is satisfied the length of the radius and the coordinates of C, the centre, are easily determined.

In § 4.12 it will be shown that this solution, with certain restrictions, gives rise to a maximum value for the enclosed area.

4.4. Examples 2 and 3

EXAMPLE 2. A heavy uniform flexible chain of given length hangs in equilibrium under gravity with its ends attached to two fixed points A and B. Find the equation of its curve.

According to statical theory the chain must hang in a vertical plane so that its potential energy is a minimum, and this principle will form the basis of our investigation.

Taking the vertical plane in which the chain hangs as the xy plane and any convenient horizontal line as the x-axis, let ρ be the mass per unit length of chain and let a and b be the abscissae of the points A and B respectively. Then the potential energy of an element of length ds at a height y above the x-axis is $\rho gy\, ds$ and so we must minimize the integral

$$I = \rho g \int\limits_A^B y\, ds, \tag{1}$$

subject to the condition that

$$l = \int\limits_A^B ds. \tag{2}$$

On transforming to the variable x it then appears that we must minimize the integral

$$I/\rho g = \int\limits_a^b y(1+y_1^2)^{\frac{1}{2}}\, dx, \tag{3}$$

subject to the condition that†

$$l = \int\limits_a^b (1+y_1^2)^{\frac{1}{2}}\, dx. \tag{4}$$

From (12), § 4.2, it follows that y must satisfy the differential equation

$$(1+y_1^2)^{\frac{1}{2}} - \frac{d}{dx}\left\{\frac{(y-\lambda)y_1}{(1+y_1^2)^{\frac{1}{2}}}\right\} = 0, \tag{5}$$

† The positive values of the roots are to be taken throughout.

where λ is a constant. It is then easy to verify that (5) is satisfied by the catenary whose equation is

$$y = \lambda + c\cosh\left(\frac{x+d}{c}\right), \tag{6}$$

where c and d are arbitrary constants.

Before proceeding further it is interesting to compare this result with that obtained in the very similar problem discussed in § 1.7. In § 1.7 the chain was allowed to pass over two smooth pegs at A and B, there was no restriction in the length, and in the solution there were only two arbitrary constants. In the solution above, given by equation (6), there are three arbitrary constants, λ, c, and d, so that the condition (2) is counterbalanced by the appearance of an extra arbitrary constant.

To complete the solution of the problem, suppose for simplicity that the points A and B have coordinates $(-a, h)$ and (a, h) respectively. Then from (6) we have

$$h - \lambda = c\cosh\left(\frac{-a+d}{c}\right) = c\cosh\left(\frac{a+d}{c}\right), \tag{7}$$

and so $d = 0$. From (6) and (4) we have

$$l = 2c\sinh\left(\frac{a}{c}\right), \tag{8}$$

which is an equation for c, since l and a are known. Now write $a = cX$ and consider the curves $2aY = lX$ and $Y = \sinh X$, where X and Y are the running coordinates. Since the curve $Y = \sinh X$ has a point of inflexion at the origin, whose tangent is inclined at an angle $\pi/4$ to the x-axis, it is easily proved that if $l \geqslant 2a$ equation (8) has only one positive root for c, and that if $l < 2a$ there are no real roots for c. In practice the first inequality is satisfied, since the length of the chain must exceed the length of AB, and (8) gives us a unique value for c. From (7) λ is then easily determined and so a unique catenary can be found for which the potential energy is stationary.

The discussion of the second variation is too elaborate for us to consider here. Some remarks will be made in § 4.12, which will show that I admits a minimum, but it is evident from physical considerations that I must be a minimum.

EXAMPLE 3. The arc Γ of a plane curve of given length l has its ends attached to two points A and B. If the arc is rotated through four right angles about an axis in its plane, determine the form of Γ in order that the superficial area so generated should be a maximum or minimum.

(This is mathematically identical with example 2.)

4.5. Example 4

A and B are two adjacent points of intersection of a plane curve Γ with the x-axis, the curve being such that any line parallel to the y-axis and having A and B on opposite sides of it cuts it in one point only. The arc of Γ lying between A and B is rotated through four right angles about the x-axis and generates a closed surface whose superficial area is Δ and whose interior volume is V. Given Δ, find the form of Γ in order that V should be a maximum or a minimum.

Here we have
$$V = \pi \int_a^b y^2\,dx, \tag{1}$$
subject to the condition that
$$\Delta = 2\pi \int_a^b y(1+y_1^2)^{\frac{1}{2}}\,dx, \tag{2}$$
where Δ is given.

From (12), § 4.2, V is stationary if y satisfies the differential equation
$$2\pi y - \lambda 2\pi(1+y_1^2)^{\frac{1}{2}} + \lambda \frac{d}{dx}\left\{\frac{2\pi y y_1}{(1+y_1^2)^{\frac{1}{2}}}\right\} = 0, \tag{3}$$
where λ is a constant. It is not difficult to verify by direct differentiation that a first integral of (3) is
$$y^2 - \frac{2\lambda y}{(1+y_1^2)^{\frac{1}{2}}} = c, \tag{4}$$
where c is an arbitrary constant (this is a special case of theorem 2, § 1.4). The surface generated by rotating the curve (4) about

the x-axis is a minimal surface,† see § 3.11, and is of great interest in surface-tension theory.‡

Since the complete integral of (4) involves elliptic functions we shall confine ourselves to the simple case where $c = 0$. In this case (4) reduces to

$$y = \frac{2\lambda}{(1+y_1^2)^{\frac{1}{2}}} = 2\lambda \cos\psi, \tag{5}$$

where ψ is the angle between the x-axis and the tangent to Γ. On differentiating (5) with respect to s, the length of arc, we have

$$\sin\psi = \frac{dy}{ds} = -2\lambda \sin\psi \frac{d\psi}{ds}. \tag{6}$$

It therefore follows that the radius of curvature is equal to -2λ and is constant, since λ is constant. Hence Γ must be an arc of a circle. If the arc is concave to the x-axis its radius of curvature will be negative and λ will then be positive.

It follows also from (5) that at A and B, where $y = 0$, we have $\psi = \frac{1}{2}\pi$, i.e. the curve must cut the x-axis at right angles and Γ is therefore a semicircle.

In § 4.12 we shall show that this case is not merely stationary but leads to a maximum. Thus among surfaces of revolution of given superficial area the one which encloses a maximum volume is the sphere.

4.6. Further isoperimetrical problems

We shall use the symbol $\oint$ to denote an integral taken once round a closed contour. The problem of this section is to find the functional form of y which renders the integral

$$I = \oint F(x, y, y_1)\, dx \tag{1}$$

a maximum or a minimum subject to the condition that

$$J = \oint \phi(x, y, y_1)\, dx. \tag{2}$$

Here J is a given constant, the two contours of integration are the same, and F and ϕ are known functional forms. An

† The sum of the principal curvatures at a point of the surface is zero. The enclosed volume of revolution is a maximum. See Forsyth, loc. cit., p. 416.

‡ See H. Lamb, *Statics*, chap. xiv.

illustration is the problem of finding the closed plane curve of given length which encloses a maximum area.

The integrals in (1) and (2) can be transformed to the more familiar forms with given end points by the introduction of parameters. Although on traversing a closed contour once, the coordinates of a point return to their original values the parameter of the representation need not do so. For example, if a point P describes the circle $x^2+y^2 = a^2$ once, its coordinates will return to their original values. But if these coordinates, (x, y) say, are given in parametric form by the equations

$$x = a\cos t, \qquad y = a\sin t,$$

then as P describes the circle once, the parameter t will vary from t_0 to $t_0+2\pi$.

Suppose that x and y, the coordinates of any point on the contour of integration, can be expressed parametrically in the form $x = p(t)$, $y = q(t)$, and let†

$$F(x,y,y_1) = F(x,y,\dot{y}/\dot{x}) = G(x,y,\dot{x},\dot{y})\,dt/dx \tag{3}$$

$$\phi(x,y,y_1) = \phi(x,y,\dot{y}/\dot{x}) = \psi(x,y,\dot{x},\dot{y})\,dt/dx, \tag{4}$$

where, as is usual in dynamics, a dot denotes differentiation with respect to t. Suppose also that when the contour is traversed completely once, the parameter t varies continuously from t_1 to t_2. On substituting in (1) and (2), taking t as the independent variable and x and y as the dependent variables, we reduce the problem to the following form. Find x and y as functions of t so as to make the integral

$$I = \int\limits_{t_1}^{t_2} G(x,y,\dot{x},\dot{y})\,dt, \tag{5}$$

a maximum or a minimum, subject to the condition that

$$J = \int\limits_{t_1}^{t_2} \psi(x,y,\dot{x},\dot{y})\,dt, \tag{6}$$

where J is a given constant.

The methods used in § 3.2 to deal with integrals similar to (5) can now be employed here. Let $x = s_1(t)$ and $y = s_2(t)$ be the

† See § 9.8 for some properties of the function $G(x, y, \dot{x}, \dot{y})$.

functional forms of x and y which make I stationary, and let us consider a small variation δI in the value of I when x and y are replaced by $x = s_1(t)+\epsilon_1 u_1(t)$ $(= s_1+\epsilon_1 u_1$ for brevity) and $y = s_2(t)+\epsilon_2 u_2(t)$ $(= s_2+\epsilon_2 u_2)$. The quantities ϵ_1 and ϵ_2 are arbitrary constants and $u_1(t)$ and $u_2(t)$ are functions of t one of which can be chosen arbitrarily (it is evident from (6) that only three of the quantities ϵ_1, ϵ_2, u_1, u_2 can be arbitrarily chosen). Since the curve is closed the parameters t_1 and t_2 correspond to the same point. We may without loss of generality assume that this point is fixed, giving us the conditions

$$u_1(t_1) = u_2(t_1) = u_1(t_2) = u_2(t_2) = 0.$$

We confine ourselves to the case where $\epsilon_1 \dot{u}_1$ and $\epsilon_2 \dot{u}_2$ both tend to zero as ϵ_1 and ϵ_2 tend to zero.

Assuming that for sufficiently small values of ϵ_1 and ϵ_2 we may employ the mean-value theorem then, from (5) we have

$$\delta I = \epsilon_1 \int\limits_{t_1}^{t_2} \left\{u_1 \frac{\partial G}{\partial s_1} + \dot{u}_1 \frac{\partial G}{\partial \dot{s}_1}\right\} dt + \epsilon_2 \int\limits_{t_1}^{t_2} \left\{u_2 \frac{\partial G}{\partial s_2} + \dot{u}_2 \frac{\partial G}{\partial \dot{s}_2}\right\} dt + O(\epsilon^2), \tag{7}$$

where $O(\epsilon^2)$ denotes terms involving ϵ_1^2, $\epsilon_1 \epsilon_2$, ϵ_1^2, and higher powers of the epsilons. On integrating by parts, as in § 1.4, and noting that u_1 vanishes at both limits of integration, we have

$$\int\limits_{t_1}^{t_2} \dot{u}_1 \frac{\partial G}{\partial \dot{s}_1} dt = u_1 \left(\frac{\partial G}{\partial \dot{s}_1}\right)_{t_1}^{t_2} - \int\limits_{t_1}^{t_2} u_1 \frac{d}{dt}\left(\frac{\partial G}{\partial \dot{s}_1}\right) dt = - \int\limits_{t_1}^{t_2} u_1 \frac{d}{dt}\left(\frac{\partial G}{\partial \dot{s}_1}\right) dt. \tag{8}$$

On combining this with the analogous result for the second integral of (7) we obtain

$$\delta I = \epsilon_1 \int\limits_{t_1}^{t_2} u_1 \left\{\frac{\partial G}{\partial s_1} - \frac{d}{dt}\left(\frac{\partial G}{\partial \dot{s}_1}\right)\right\} dt + \epsilon_2 \int\limits_{t_1}^{t_2} u_2 \left\{\frac{\partial G}{\partial s_2} - \frac{d}{dt}\left(\frac{\partial G}{\partial \dot{s}_2}\right)\right\} dt + O(\epsilon^2). \tag{9}$$

Noting that J is a constant we may similarly deduce from (6) that

$$0 = \epsilon_1 \int\limits_{t_1}^{t_2} u_1 \left\{\frac{\partial \psi}{\partial s_1} - \frac{d}{dt}\left(\frac{\partial \psi}{\partial \dot{s}_1}\right)\right\} dt + \epsilon_2 \int\limits_{t_1}^{t_2} u_2 \left\{\frac{\partial \psi}{\partial s_2} - \frac{d}{dt}\left(\frac{\partial \psi}{\partial \dot{s}_2}\right)\right\} dt + O(\epsilon^2). \tag{10}$$

Equation (10) shows us once again that all four quantities ϵ_1, ϵ_2, $u_1(t)$, and $u_2(t)$ cannot be arbitrarily chosen. For we can take ϵ_1 and ϵ_2 to be arbitrary constants, $u_2(t)$ to be an arbitrary function of t, and then regard (10) as an integral equation for the determination of $u_1(t)$.

In order to find the conditions which ensure the vanishing of the first variation of I we proceed as follows. Choose $x = s_1(t)$ and $y = s_2(t)$ to satisfy the equation

$$\frac{\partial G}{\partial x}-\frac{d}{dt}\left(\frac{\partial G}{\partial \dot{x}}\right)-\lambda\left\{\frac{\partial \psi}{\partial x}-\frac{d}{dt}\left(\frac{\partial \psi}{\partial \dot{x}}\right)\right\}=0, \tag{11}$$

where λ is an undetermined constant whose value can be found by a method given later. This equation by itself is insufficient to define the two functions $s_1(t)$ and $s_2(t)$.

Now multiply (10) by λ and subtract from (9). From (11) the co-factor of $u_1(t)$ vanishes and we have

$$\delta I = \epsilon_2 \int_{t_1}^{t_2} u_2(t)\left[\left\{\frac{\partial G}{\partial s_2}-\frac{d}{dt}\left(\frac{\partial G}{\partial \dot{s}_2}\right)\right\}-\lambda\left\{\frac{\partial \psi}{\partial s_2}-\frac{d}{dt}\left(\frac{\partial \psi}{\partial \dot{s}_2}\right)\right\}\right]dt+O(\epsilon^2). \tag{12}$$

If I is to be stationary the coefficient of ϵ_2 must vanish. Since $u_2(t)$ is an arbitrary function of t, the arguments of § 1.4 then enable us to deduce that $x = s_1(t)$ and $y = s_2(t)$ must also satisfy

$$\frac{\partial G}{\partial y}-\frac{d}{dt}\left(\frac{\partial G}{\partial \dot{y}}\right)-\lambda\left\{\frac{\partial \psi}{\partial y}-\frac{d}{dt}\left(\frac{\partial \psi}{\partial \dot{y}}\right)\right\}=0. \tag{13}$$

Equations (11) and (13) suffice for the determination of both $s_1(t)$ and $s_2(t)$. The arbitrary constants enable us to choose those solutions of (11) and (13) which satisfy given conditions, such as passing through given points, and the extra undetermined constant λ can then be chosen so that condition (2) is also satisfied.

We have therefore proved:

THEOREM 12. *By suitable choice of the arbitrary constants and of the constant* λ *it is possible to find solutions of* (11) *and* (13) *which make the integral of equation* (1) *stationary and at the same time satisfy condition* (2).

This result is illustrated by the following example.

4.7. Example 5

Find the plane curve of given length which encloses a maximum area.

Let I denote the area and L the length of the curve and let the coordinates (x,y) of any point on the curve be expressed as functions of the parameter t. Then†

$$I = \tfrac{1}{2}\int_{t_1}^{t_2} (x\dot{y}-\dot{x}y)\,dt \tag{1}$$

and

$$L = \int_{t_1}^{t_2} (\dot{x}^2+\dot{y}^2)^{\frac{1}{2}}\,dt, \tag{2}$$

where the positive value of the root is to be taken.

Equations (11) and (13) of § 4.6 become

$$\tfrac{1}{2}\dot{y}+\tfrac{1}{2}\frac{d}{dt}(y)-\lambda\left[-\frac{d}{dt}\left\{\frac{\dot{x}}{(\dot{x}^2+\dot{y}^2)^{\frac{1}{2}}}\right\}\right] = 0, \tag{3}$$

and

$$-\tfrac{1}{2}\dot{x}-\tfrac{1}{2}\frac{d}{dt}(x)-\lambda\left[-\frac{d}{dt}\left\{\frac{\dot{y}}{(\dot{x}^2+\dot{y}^2)^{\frac{1}{2}}}\right\}\right] = 0. \tag{4}$$

If ds is the element of arc, these equations are easily reduced to

$$\frac{dy}{ds}+\lambda\frac{d^2x}{ds^2} = 0 \tag{5}$$

and

$$-\frac{dx}{ds}+\lambda\frac{d^2y}{ds^2} = 0. \tag{6}$$

On solving these simultaneous equations for x and y in terms of s we obtain

$$x = a+c\sin\left(\frac{s}{\lambda}+\alpha\right) \tag{7}$$

and

$$y = b-c\cos\left(\frac{s}{\lambda}+\alpha\right), \tag{8}$$

where a, b, c, and α are arbitrary constants. This solution represents a circle of radius c and centre (a,b). Evidently $c = L/2\pi$. Substituting from (7) and (8) in (2) and taking the s-limits of integration to be s_1 and $s_1+2\pi c$ we deduce that

$$\lambda = c = L/2\pi.$$

† Courant, *Differential and Integral Calculus*, vol. i, p. 273.

The remaining arbitrary constants a, b, and α can be determined if we are given one point on the extremal and the slope of the tangent there.

An alternative method is to note that if ψ is the angle between the x-axis and the tangent to the curve at the point (x, y), then $dx/ds = \cos\psi$, $dy/ds = \sin\psi$, so that (6) becomes

$$\cos\psi - \lambda\,(d\psi/ds)\cos\psi = 0.$$

Thus, since λ is constant, the radius of curvature must be constant at all points and equal to λ. Equations (5) and (6) can be easily integrated in terms of ψ. The results are

$$x = a+\lambda\sin(\psi+\alpha), \tag{9}$$

$$y = b-\lambda\cos(\psi+\alpha), \tag{10}$$

where a, b, α are arbitrary constants whose values can be determined as above.

In § 4.13 it will be shown that this solution gives a maximum value to the integral I of (1). Discontinuous solutions are possible in some cases.† For example, if the contour is required to pass through three non-collinear points A, B, C, then we must find three circular arcs of equal radius whose lengths add up to the given value L, each arc terminating in two of the points A, B, C. If $BC = a$, $CA = b$, and $AB = c$, then the angle subtended by BC at the centre of arc BC is $2\sin^{-1}(a/2\lambda)$, etc., where λ is the common radius. Hence equation (2) becomes

$$2\lambda\sin^{-1}\left(\frac{a}{2\lambda}\right)+2\lambda\sin^{-1}\left(\frac{b}{2\lambda}\right)+2\lambda\sin^{-1}\left(\frac{c}{2\lambda}\right) = L. \tag{11}$$

This is an equation for λ, and if it can be solved the positions of the centres and the arcs are then easily obtained.

If the three points are collinear, on taking $AB = c$, $BC = a$, and B to lie between A and C the equation (2) for λ becomes

$$2\lambda\sin^{-1}\left(\frac{a}{2\lambda}\right)+2\lambda\sin^{-1}\left(\frac{c}{2\lambda}\right)+2\lambda\sin^{-1}\left(\frac{a+c}{2\lambda}\right) = L. \tag{12}$$

† The observations of § 1.17 apply to these cases.

4.8. Subsidiary equations of non-integral type

In previous sections relative maxima and minima have been obtained when the subsidiary equation involves integration. In this section we shall deal with cases where the subsidiary equation takes other forms. As an illustration consider the problem of finding geodesics on a surface. This requires us to minimize an integral of the type

$$I = \int_{t_1}^{t_2} (\dot{x}^2+\dot{y}^2+\dot{z}^2)^{\frac{1}{2}}\, dt, \tag{1}$$

where the point (x, y, z) is restricted to lie on the surface whose equation is

$$S(x, y, z) = 0. \tag{2}$$

If (2) could be solved for one of the variables, for example z, it would be possible to eliminate z from (1) and reduce the problem to one of unrestricted maxima and minima with one independent and one dependent variable. In practice the elimination may be too difficult or too inconvenient to perform and in some cases a certain desirable symmetry between the variables may be lost in the process. It is therefore worth while investigating the problem in the form stated above.

We shall consider a more general problem than the one just mentioned. Find the maximum and minimum values of the integral

$$I = \int_{t_1}^{t_2} G(x, y, t, \dot{x}, \dot{y})\, dt, \tag{3}$$

where the variables are subject to the condition that

$$S(x, y, t, \dot{x}, \dot{y}) = 0, \tag{4}$$

G and S being given functional forms.

As in § 4.6, let $x = s_1(t) = s_1$ and $y = s_2(t) = s_2$ be the functional forms of x and y which make I stationary. Consider δI, the variation in I when, for a given value of t, the values of x and y are varied to $x = s_1+\epsilon_1 u_1$ and $y = s_2+\epsilon_2 u_2$. Here ϵ_1 and ϵ_2 are arbitrary constants and u_2 $(= u_2(t))$ is an arbitrary function of t, the fourth quantity, u_1 $(= u_1(t))$ being dependent upon the three previous ones. At the end points it is assumed, for simplicity, that $u_1(t_1) = u_2(t_1) = u_1(t_2) = u_2(t_2) = 0$.

On employing the same analysis for (3) as for (5), § 4.6, we may deduce that

$$\delta I = \epsilon_1 \int_{t_1}^{t_2} u_1\left\{\frac{\partial G}{\partial s_1} - \frac{d}{dt}\left(\frac{\partial G}{\partial \dot{s}_1}\right)\right\} dt + \epsilon_2 \int_{t_1}^{t_2} u_2\left\{\frac{\partial G}{\partial s_2} - \frac{d}{dt}\left(\frac{\partial G}{\partial \dot{s}_2}\right)\right\} dt + O(\epsilon^2), \tag{5}$$

where $O(\epsilon^2)$ denotes terms involving ϵ_1^2, $\epsilon_1\epsilon_2$, ϵ_2^2 and higher powers of the epsilons.

From equation (4) we also have

$$\epsilon_1 u_1 \frac{\partial S}{\partial s_1} + \epsilon_1 \dot{u}_1 \frac{\partial S}{\partial \dot{s}_1} + \epsilon_2 u_2 \frac{\partial S}{\partial s_2} + \epsilon_2 \dot{u}_2 \frac{\partial S}{\partial \dot{s}_2} + O(\epsilon^2) = 0. \tag{6}$$

Now if μ is any function of x, y, and t, by straightforward differentiation of the terms on the right-hand side it is easily verified that

$$\mu\left(\epsilon_1 \dot{u}_1 \frac{\partial S}{\partial \dot{s}_1} + \epsilon_2 \dot{u}_2 \frac{\partial S}{\partial \dot{s}_2}\right) = \frac{d}{dt}\left\{\mu\left(\epsilon_1 u_1 \frac{\partial S}{\partial \dot{s}_1} + \epsilon_2 u_2 \frac{\partial S}{\partial \dot{s}_2}\right)\right\} - $$
$$- \left\{\epsilon_1 u_1 \frac{d}{dt}\left(\mu \frac{\partial S}{\partial \dot{s}_1}\right) + \epsilon_2 u_2 \frac{d}{dt}\left(\mu \frac{\partial S}{\partial \dot{s}_2}\right)\right\}. \tag{7}$$

Hence, on multiplying (6) by μ and subtracting (7) we have

$$\epsilon_1 u_1\left\{\mu \frac{\partial S}{\partial s_1} - \frac{d}{dt}\left(\mu \frac{\partial S}{\partial \dot{s}_1}\right)\right\} + \epsilon_2 u_2\left\{\mu \frac{\partial S}{\partial s_2} - \frac{d}{dt}\left(\mu \frac{\partial S}{\partial \dot{s}_2}\right)\right\} + $$
$$+ \frac{d}{dt}\left\{\mu\left(\epsilon_1 u_1 \frac{\partial S}{\partial \dot{s}_1} + \epsilon_2 u_2 \frac{\partial S}{\partial \dot{s}_2}\right)\right\} + O(\epsilon^2) = 0. \tag{8}$$

Integrating, and noting that

$$u_1(t_1) = u_2(t_1) = u_1(t_2) = u_2(t_2) = 0,$$

we have

$$\epsilon_1 \int_{t_1}^{t_2} u_1\left\{\mu \frac{\partial S}{\partial s_1} - \frac{d}{dt}\left(\mu \frac{\partial S}{\partial \dot{s}_1}\right)\right\} dt + \epsilon_2 \int_{t_1}^{t_2} u_2\left\{\mu \frac{\partial S}{\partial s_2} - \frac{d}{dt}\left(\mu \frac{\partial S}{\partial \dot{s}_2}\right)\right\} dt + $$
$$+ O(\epsilon^2) = 0. \tag{9}$$

Now assume that μ is chosen so that $x = s_1(t)$ and $y = s_2(t)$ satisfy the partial differential equation

$$\frac{\partial G}{\partial x} - \frac{d}{dt}\left(\frac{\partial G}{\partial \dot{x}}\right) - \left\{\mu \frac{\partial S}{\partial x} - \frac{d}{dt}\left(\mu \frac{\partial S}{\partial \dot{x}}\right)\right\} = 0. \tag{10}$$

On subtracting equation (9) from (5) it then follows that the cofactor of u_1 vanishes, so that

$$\delta I = \epsilon_2 \int_{t_1}^{t_2} u_2 \left[\frac{\partial G}{\partial s_2} - \frac{d}{dt}\left(\frac{\partial G}{\partial \dot{s}_2}\right) - \left\{\mu \frac{\partial S}{\partial s_2} - \frac{d}{dt}\left(\mu \frac{\partial S}{\partial \dot{s}_2}\right)\right\}\right] dt + O(\epsilon^2). \tag{11}$$

If I is a maximum or minimum, then the coefficient of ϵ_2 must vanish. Therefore, since $u_2(t)$ is an arbitrary function of t, the arguments of § 1.4 prove that the functions $x = s_1(t)$ and $y = s_2(t)$ must satisfy the following equation:

$$\frac{\partial G}{\partial y} - \frac{d}{dt}\left(\frac{\partial G}{\partial \dot{y}}\right) - \left\{\mu \frac{\partial S}{\partial y} - \frac{d}{dt}\left(\mu \frac{\partial S}{\partial \dot{y}}\right)\right\} = 0. \tag{12}$$

The three equations (4), (10), and (12) suffice to determine x, y, and μ as functions of t.

We have thus proved:

THEOREM 13. *Solutions of the simultaneous equations* (10) *and* (12) *can be found which satisfy equation* (4) *and which make the integral I of* (3) *stationary.*

If another dependent variable, e.g. z, appears in (3) and (4) in addition to x and y we still use the same arguments. It is then found that for I to be stationary a further characteristic equation is required in addition to (10) and (12), one in which the partial derivatives are taken with respect to z instead of with respect to x or y. The factor μ is the same for (10), (12), and the additional equation.

The details of proof are left to the reader.

4.9. Example 6. Geodesics

To find the geodesics on a given surface.

Let
$$S(x, y, z) = 0 \tag{1}$$

be the equation of the given surface and assume that the coordinates of a point situated on a curve which lies on (1) can be expressed as functions of a parameter t. Then the length of arc joining the two points A and B is I where

$$I = \int_{t_1}^{t_2} (\dot{x}^2+\dot{y}^2+\dot{z}^2)^{\frac{1}{2}}\,dt, \tag{2}$$

and the problem reduces to that of finding the minimum value of I subject to the restriction imposed by equation (1).

Equation (10), § 4.8, becomes

$$-\frac{d}{dt}\left\{\frac{\dot{x}}{(\dot{x}^2+\dot{y}^2+\dot{z}^2)^{\frac{1}{2}}}\right\}-\mu\frac{\partial S}{\partial x} = 0, \tag{3}$$

and in addition there are two corresponding equations for y and z. If s denotes the length of arc, (3) simplifies to

$$\frac{ds}{dt}\frac{d^2x}{ds^2}+\mu\frac{\partial S}{\partial x} = 0 \tag{4}$$

together with two corresponding results for y and z. It follows that

$$\frac{d^2x/ds^2}{\partial S/\partial x} = \frac{d^2y/ds^2}{\partial S/\partial y} = \frac{d^2z/ds^2}{\partial S/\partial z} = -\frac{\mu}{ds/dt}. \tag{5}$$

The first three fractions have their numerators proportional to the direction cosines of the principal normal to the geodesic† and their denominators proportional to the direction cosines of the normal to the surface (1). The principal normal to the geodesic must therefore coincide with the normal to the surface.

From the four equations given by (5) together with (1) it is possible to determine x, y, z, and μ as functions of t. A slight simplification is effected by taking the parameter t to be equal to s, the length of arc, so that $ds/dt = 1$.

4.10. Examples 7–9. Geodesics on a sphere

EXAMPLE 7. We illustrate the theory of § 4.9 by finding the geodesics on a sphere, a problem already discussed in § 1.10.

Taking the centre of the sphere at the origin its equation is

$$x^2+y^2+z^2 = a^2, \tag{1}$$

† C. Smith, *Solid Geometry*, p. 202, § 233.

and taking the parameter t to be equal to s, the length of arc, equations (5), § 4.9, become

$$\frac{1}{x}\frac{d^2x}{ds^2} = \frac{1}{y}\frac{d^2y}{ds^2} = \frac{1}{z}\frac{d^2z}{ds^2} = -2\mu. \tag{2}$$

But on differentiating (1) we have

$$x\frac{dx}{ds}+y\frac{dy}{ds}+z\frac{dz}{ds} = 0, \tag{3}$$

and on differentiating further we have

$$x\frac{d^2x}{ds^2}+y\frac{d^2y}{ds^2}+z\frac{d^2z}{ds^2} = -\left(\frac{dx}{ds}\right)^2-\left(\frac{dy}{ds}\right)^2-\left(\frac{dz}{ds}\right)^2 = -1. \tag{4}$$

Hence from (2) and (4) we obtain

$$-2\mu(x^2+y^2+z^2) = -1, \tag{5}$$

and so from (1) it follows that

$$\mu = \frac{1}{2a^2}. \tag{6}$$

It should be noted that in general μ is a function of x, y, z, and not, as here, a constant. Equations (2) can now be simplified to

$$\frac{d^2x}{ds^2}+\frac{x}{a^2} = 0, \quad \frac{d^2y}{ds^2}+\frac{y}{a^2} = 0, \quad \frac{d^2z}{ds^2}+\frac{z}{a^2} = 0, \tag{7}$$

the solutions of which are

$$x = A\cos\left(\frac{s}{a}+\alpha\right), \quad y = B\cos\left(\frac{s}{a}+\beta\right), \quad \text{and} \quad z = C\cos\left(\frac{s}{a}+\gamma\right). \tag{8}$$

Here the constants A, B, C, α, β, γ are related in such a manner that only three are arbitrary. For on inserting the results of (8) in (1) we have an expression which is identically true for all values of s. We may then equate the coefficients of $\cos(2s/a)$, $\sin(2s/a)$ and the term independent of s to zero, and so obtain three relations between the six constants.

Eliminating s from equations (8) we deduce that

$$\begin{vmatrix} x & A\cos\alpha & A\sin\alpha \\ y & B\cos\beta & B\sin\beta \\ z & C\cos\gamma & C\sin\gamma \end{vmatrix} = 0. \tag{9}$$

Hence the geodesics must lie on a plane through the centre and so must lie along arcs of great circles. This agrees with the result obtained in § 1.10.

EXAMPLE 8. Given the cylindrical surface whose equation is $x^2+y^2 = a^2$ and two points A and B whose coordinates are $(a, 0, 0)$ and $(a\cos\alpha, a\sin\alpha, b)$ respectively. Prove that the geodesics through A and B are given by

$$x = a\cos t, \quad y = a\sin t, \quad z = bt/(2n\pi+\alpha),$$

where n is an integer.

EXAMPLE 9. Given a surface of revolution whose axis lies along the z-axis. Let the Cartesian and cylindrical coordinates of a point P on the surface be (x, y, z) and (r, θ, z) respectively. Prove that along a geodesic $x\dfrac{dy}{ds}-y\dfrac{dx}{ds} = c$ and $r^2\dfrac{d\theta}{ds} = c$, where c is a constant, the same for both equations.

4.11. Non-holonomic dynamical constraints

We now deal with some problems of relative maxima and minima which are of great importance in dynamics in connexion with systems of forces known as non-holonomic dynamical systems. They will occur again in the next chapter when we deal with Hamilton's principle.

Non-holonomic restraints occur when there is one independent and several dependent variables and they are significant only when there are at least three dependent variables. For simplicity we shall confine ourselves to the case of three dependent variables, but we shall employ arguments which can be applied to the general case.

If t is the independent and x, y, z the three dependent variables, then a dynamical constraint is given by an equation of the form

$$P\dot{x}+Q\dot{y}+R\dot{z} = 0, \tag{1 a}$$

or what is equivalent,

$$P\,dx+Q\,dy+R\,dz = 0, \tag{1}$$

where P, Q, R are functions of x, y, z, and t. Such equations are not, in general, integrable and can only be integrated if P, Q, R satisfy the integrability condition†

$$P\left(\frac{\partial Q}{\partial z}-\frac{\partial R}{\partial y}\right)+Q\left(\frac{\partial R}{\partial x}-\frac{\partial P}{\partial z}\right)+R\left(\frac{\partial P}{\partial y}-\frac{\partial Q}{\partial x}\right)=0. \tag{2}$$

If this condition is satisfied (1) is integrable. We can then solve for z and by elimination reduce the problem to one with two variables and with no condition of restraint. If (2) is *not* satisfied, and (1) is therefore *not integrable*, then the problem cannot be reduced in this way. We then have a condition which corresponds to the case of a non-holonomic dynamical system.

We now assume that (2) is not satisfied, and therefore (1) is not integrable, and proceed to find the conditions which ensure that the integral

$$I=\int\limits_{t_1}^{t_2} G(x,y,z,t,\dot{x},\dot{y},\dot{z})\,dt \tag{3}$$

is stationary subject to condition (1 a) or (1). In the stationary case let $x = s_1(t)$, $y = s_2(t)$, $z = s_3(t)$, denoted for brevity by s_1, s_2, s_3 respectively, and in the varied case for a given value of t let x, y, z change from these values to $s_1+\epsilon_1 u_1$, $s_2+\epsilon_2 u_2$, $s_3+\epsilon_3 u_3$ respectively. Here ϵ_1, ϵ_2, and ϵ_3 are arbitrary constants and u_2 and u_3 are arbitrary functions of t. The remaining quantity, u_1, which is also a function of t, is expressible in terms of the other five by virtue of (1 a). At the end points, $t = t_1$ and $t = t_2$, the functions u_1, u_2, and u_3 all vanish.

In (1) we may write $dx = \epsilon_1 u_1$, $dy = \epsilon_2 u_2$, $dz = \epsilon_3 u_3$, multiply by an undetermined factor ν, and integrate with respect to t. We then have

$$\epsilon_1\int\limits_{t_1}^{t_2}\nu\, Pu_1\,dt+\epsilon_2\int\limits_{t_1}^{t_2}\nu\, Qu_2\,dt+\epsilon_3\int\limits_{t_1}^{t_2}\nu\, Ru_3\,dt=0. \tag{4}$$

We can now proceed as in § 4.8, replacing equation (9), § 4.8, by (4). First choose $x = s_1(t)$, $y = s_2(t)$, $z = s_3(t)$, and ν to satisfy the equation

$$\frac{\partial G}{\partial x}-\frac{d}{dt}\left(\frac{\partial G}{\partial \dot{x}}\right)-\nu P=0. \tag{5}$$

† Forsyth, *Differential Equations*, 4th edition, p. 309; Piaggio, *Differential Equations* (1928), p. 137.

Then write $x = s_1+\epsilon_1 u_1$, etc., in (3). On calculating the value of δI it is evident that the result must be the same as (5), § 4.8, except that we have a third integral to allow for owing to the presence of the additional variable z. Subtracting (4) from δI the coefficient of ϵ_1 vanishes, owing to (5), and we are left with

$$\delta I = \epsilon_2 \int\limits_{t_1}^{t_2} u_2\left\{\frac{\partial G}{\partial s_2}-\frac{d}{dt}\left(\frac{\partial G}{\partial \dot{s}_2}\right)-\nu Q\right\}dt+$$

$$+\epsilon_3 \int\limits_{t_1}^{t_2} u_3\left\{\frac{\partial G}{\partial s_3}-\frac{d}{dt}\left(\frac{\partial G}{\partial \dot{s}_3}\right)-\nu R\right\}dt+O(\epsilon^2). \tag{6}$$

If I is to be stationary, then the coefficients of ϵ_2 and ϵ_3 must vanish. Since u_2 and u_3 are arbitrary functions of t, the arguments of § 1.4 enable us to deduce further that $x = s_1$, $y = s_2$, and $z = s_3$ must also satisfy the equations

$$\frac{\partial G}{\partial y}-\frac{d}{dt}\left(\frac{\partial G}{\partial \dot{y}}\right)-\nu Q = 0 \tag{7}$$

and

$$\frac{\partial G}{\partial z}-\frac{d}{dt}\left(\frac{\partial G}{\partial \dot{z}}\right)-\nu R = 0 \tag{8}$$

in addition to (5).

The three characteristic equations (5), (7), and (8) together with (1) suffice to determine x, y, z, and the factor ν.

These ideas can be generalized to the case when n dependent variables $x_1, x_2,..., x_n$ appear in the integrand of I. When there are conditional equations of the type (4), § 4.8, and also non-holonomic restraints of the type (1) above, the general result is as follows:

For a stationary value of

$$I = \int\limits_{t_1}^{t_2} G(x_1, x_2,..., x_n; \dot{x}_1, \dot{x}_2,..., \dot{x}_n; t)\, dt, \tag{9}$$

subject to conditional equations of the type

$$S_m(x_1, x_2,..., x_n; \dot{x}_1, \dot{x}_2,..., \dot{x}_n; t) = 0, \tag{10}$$

$m = 1, 2,..., p$, and non-holonomic equations of constraint of the type

$$P_{i,1}\, dx_1+P_{i,2}\, dx_2+...+P_{i,n}\, dx_n = 0, \tag{11}$$

$i = 1, 2,..., q$, the dependent variables must satisfy the n characteristic equations

$$\frac{\partial G}{\partial x_r} - \frac{d}{dt}\left(\frac{\partial G}{\partial \dot{x}_r}\right) - \sum_{m=1}^{p} \left\{\mu_m \frac{\partial S_m}{\partial x_r} - \frac{d}{dt}\left(\mu_m \frac{\partial S_m}{\partial \dot{x}_r}\right)\right\} - \sum_{i=1}^{q} \nu_i P_{i,r} = 0, \quad (12)$$

$r = 1, 2,..., n$, where μ_m and ν_i are independent of r.

The equations (10), (11), and (12) suffice to determine all the quantities $x_1, x_2,..., x_n$ together with the multipliers μ_m and ν_i.

4.12. The second variation

The conditions obtained so far ensure that the integrals concerned are stationary. In order to determine whether these integrals are maxima or minima a study must be made of the second variation, i.e. those terms in δI which contain ϵ_1^2, $\epsilon_1 \epsilon_2$, ϵ_2^2, etc., as factors. The ideas which underlie this study are the same as those expounded in Chapter II. But the nature of the analysis is in general so elaborate and intricate that we shall confine ourselves to a statement of the results, either without proof or with just an outline of one.

From the characteristic equations we can derive accessory equations which enable us to express the second variations in forms whose signs are easily determined. The theory is analogous to that of § 2.4, but far more intricate. It is then possible to deduce two tests which are generalizations of the Legendre test of §§ 1.5 and 2.5 and the Jacobi test of Chapter II. The first test requires the sign of certain derivatives to remain constant throughout the range of integration and the second defines the maximum permissible length of this range.

The second test consists of an application of Jacobi's theory of conjugate points outlined in Chapter II. Given a point on a curve (or a curve on a surface) it is possible to generalize Jacobi's theory and define conjugate points (or curves). The second test then requires that the range (domain) of integration must lie within the arc (region) bounded by a point (curve) and its next conjugate point (conjugate curve). The ideas can be extended to integrals with any number of variables of integration.

Conjugate points may be defined as in § 2.6 by considering the zeros of solutions of the accessory equations. From this definition analytical expressions, such as (6), § 2.6, or (3), § 2.12, can be obtained for the determination of all the conjugates of a given point or curve. It can also be shown, as in § 2.9, that conjugate points possess the following property. Let E denote the family of extremals passing through a given point P (or curve C), then the limiting points (curves) of intersection of neighbouring members of E are the conjugates of P (or C). Of these methods the second is, in general, much the easier to use in practice. If the family E has an envelope, then the conjugate points (curves) can be found as points of contact with the envelope, as in §§ 2.10, and 2.11.

Consider the results of § 4.2, where it was shown that solutions of the characteristic equation (10), § 4.2, make the integral I, of (1), § 4.2, stationary. In order to find the second variation we must find the coefficient of ϵ^2 in (9), § 4.2. This can be done by finding the coefficients of ϵ^2 in (5) and (6), § 4.2, multiplying the second by λ, and subtracting from the first. The second variation is then found to be

$$\tfrac{1}{2}\epsilon^2 \int\limits_a^b \left[\left\{t^2\frac{\partial^2 F}{\partial s^2}+2tt_1\frac{\partial^2 F}{\partial s\partial s_1}+t_1^2\frac{\partial^2 F}{\partial s_1^2}\right\}- -\lambda\left\{t^2\frac{\partial^2 \phi}{\partial s^2}+2tt_1\frac{\partial^2 \phi}{\partial s\partial s_1}+t_1^2\frac{\partial^2 \phi}{\partial s_1^2}\right\}\right]dx, \tag{1}$$

where $t = t(x)$ and $t_1 = dt/dx$. The arguments of § 1.5 then indicate that if the range of integration is sufficiently small the dominant term in (1) is

$$t_1^2\left\{\frac{\partial^2 F}{\partial s_1^2}-\lambda\frac{\partial^2 \phi}{\partial s_1^2}\right\}. \tag{2}$$

Reverting to the more customary y notation the following result can be proved:†

THEOREM 14. *If the range of integration of the integral I of* (1), § 4.2, *is sufficiently small, and if the expression*

$$E = \frac{\partial^2 F}{\partial y_1^2}-\lambda\frac{\partial^2 \phi}{\partial y_1^2}$$

has the same sign throughout this range, then the integral I has a maximum if E is negative and a minimum if E is positive.

† Forsyth, loc. cit., p. 407.

Consider the example of § 4.3. It was required to find the curve Γ, of given length, which joined two given points A and B in such a manner that Γ and the chord AB enclosed a maximum area. The characteristic curves, as shown in § 4.3, are circles of radius λ and the function $F-\lambda\phi$ is

$$\tfrac{1}{2}r^2-\lambda(r^2+r_1^2)^{\frac{1}{2}},$$

where (r,θ) are polar coordinates and $r_1 = dr/d\theta$. The test function of theorem 14 is

$$\frac{\partial^2 F}{\partial r_1^2}-\lambda\frac{\partial^2 \phi}{\partial r_1^2},$$

and on evaluation, this is found to be equal to $-\lambda r^2(r^2+r_1^2)^{-\frac{3}{2}}$. Hence, taking the positive value of the root, for a sufficiently small value of the length AB the area will be a maximum.

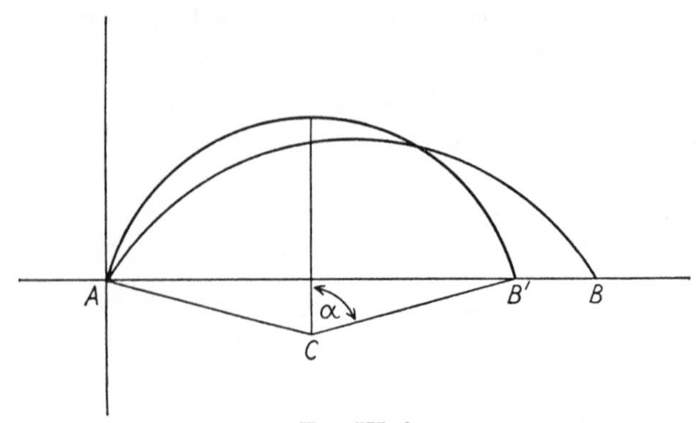

Fig. IV. 2.

To determine the permissible extent of the length AB we shall find A', the conjugate of A, by considering the intersection of neighbouring extremals through A. Take A as the origin and B on the x-axis (Fig. IV. 2) and let B' be another point on the x-axis, where $AB' < 2l$, the given length of the curve Γ. Let C be the centre of the circle which passes through A and B' whose arc $AB' = 2l$, and let 2α denote the angle subtended by AB' at C, so that its radius is l/α. Then (Fig. IV. 2) the equation of this circle is

$$x^2+y^2-\frac{2l}{\alpha}(x\sin\alpha-y\cos\alpha) = 0. \tag{3}$$

When B' coincides with B, let $\alpha = \beta$ so that $AB = (2l\sin\beta)/\beta$, an equation for β which must have one root between 0 and π if $AB < 2l$. The equation for the stationary curve Γ will then be

$$x^2+y^2-\frac{2l}{\beta}(x\sin\beta-y\cos\beta) = 0. \tag{4}$$

Regarding α as a variable parameter, (3) is the equation of a family of extremals each of which (i) satisfies the characteristic equation (4), § 4.3; (ii) passes through A; and (iii) has the length of arc between A and B equal to $2l$, the given length. Hence A', conjugate of A, can be found as the limiting point of intersection of (4) and a neighbouring circle of (3). In the usual manner put $\alpha = \beta+\delta\beta$ in (3), subtract (4), divide by $\delta\beta$, and then let $\delta\beta$ tend to zero. The result is that the required conjugate A' must lie on the line

$$x(\beta\cos\beta-\sin\beta)+y(\beta\sin\beta+\cos\beta) = 0. \tag{5}$$

Evidently, from Fig. IV. 2, if A' is to lie outside the arc AB the line (5) must make an angle with the x-axis whose tangent is negative. Hence for a maximum to be possible, the expression

$$\frac{\beta\cos\beta-\sin\beta}{\beta\sin\beta+\cos\beta} \tag{6}$$

must be positive. Since β lies between 0 and π the numerator is always negative and so the denominator must also be negative and this requires that β should lie between $160^\circ\ 20'$ (approximately) and 180°. The area enclosed by Γ and the chord AB then admits of a maximum if the arc Γ subtends an angle greater than $320^\circ\ 40'$ at the centre.

By making the length of the chord AB tend to zero it follows that a plane curve of given length encloses a maximum area when it is in the form of a circle, thus completing the solution of the isoperimetrical problem of § 4.7.

In the case of example 2, § 4.4, where it was required to find the shape in which a heavy uniform flexible chain hangs under gravity, the Legendre test depends upon the sign of

$$\frac{\partial^2 F}{\partial y_1^2}-\lambda\frac{\partial^2\phi}{\partial y_1^2}.$$

From (3) and (4), § 4.4 this is equal to $(y-\lambda)(1+y_1^2)^{-\frac{3}{2}}$ which, from (6), § 4.4, is found to be equal to

$$c\,\text{sech}^2\left(\frac{x+d}{c}\right).$$

Hence for positive c the potential energy admits of a minimum if the range of integration is sufficiently small. The Jacobi test is too intricate to discuss here, but it is obvious from physical considerations that the only limitation is that the length of the chain must exceed the distance between the points of suspension.

In example 4, § 4.5, where it was required to find the closed surface of revolution with given superficial area and maximum volumetric content, the Legendre test function

$$\frac{\partial^2 F}{\partial y_1^2}-\lambda\frac{\partial^2\phi}{\partial y_1^2}$$

is equal to $-2\pi\lambda y(1+y_1^2)^{-\frac{3}{2}}$. From (5), § 4.5, this is equal to $-4\pi\lambda^2\cos^4\psi$, so that the enclosed volume admits of a maximum for a sufficiently small range of integration. The Jacobi test is, once again, too intricate for discussion here.

4.13. Isoperimetrical problems (second variation)

In this section we shall state without proof the Legendre and Jacobi tests for the isoperimetrical problems of § 4.6. Indications will be given to show how some of the expressions occurring in the tests are obtained.

Let the solutions of the characteristic equations (11) and (13), § 4.6, be given by

$$x = X(t,a_1,a_2,\lambda) \tag{1}$$

$$y = Y(t,a_1,a_2,\lambda), \tag{2}$$

where a_1, a_2, are arbitrary constants and λ is the constant introduced in (11), § 4.6. Now conjugate points are the limiting positions of intersections of neighbouring extremals. A neighbour of the extremal defined by equations (1) and (2) is obtained by replacing a_1 by $a_1+\epsilon\alpha_1$, a_2 by $a_2+\epsilon\alpha_2$, and λ by $\lambda+\epsilon\mu$, where ϵ is small. It is then not difficult to show that the coordinates of the points of intersection must satisfy the equation

$$\alpha_1\frac{\partial(x,y)}{\partial(t,a_1)}+\alpha_2\frac{\partial(x,y)}{\partial(t,a_2)}+\mu\frac{\partial(x,y)}{\partial(t,\lambda)} = 0. \tag{3}$$

Here the usual notation for the Jacobian of two functions has been employed, namely

$$\frac{\partial(x,y)}{\partial(u,v)} = \frac{\partial x}{\partial u}\frac{\partial y}{\partial v} - \frac{\partial x}{\partial v}\frac{\partial y}{\partial u}. \tag{4}$$

Thus, if t_1 is the parameter of A and t_1' that of its conjugate point A', both must satisfy (3) giving us two equations. If we could find a third linear equation between α_1, α_2, μ it would be possible to eliminate these three quantities and so obtain the desired relation between t_1 and t_1'.

The third equation can be obtained from (10), § 4.6, by means of some rather elaborate analysis. By choosing $\epsilon_1 = \epsilon_2$ (it can be shown that this entails no loss in generality), dividing (10), § 4.6, by ϵ_1 and then letting ϵ_1 tend to zero, we get an equation which can be transformed into the following form. Letting

$$K = \frac{\partial^2\phi}{\partial y\partial\dot{x}} - \frac{\partial^2\phi}{\partial x\partial\dot{y}} + \frac{\dot{x}\ddot{y}-\ddot{x}\dot{y}}{\dot{x}\dot{y}}\frac{\partial^2\phi}{\partial\dot{x}\partial\dot{y}}, \tag{5}$$

where ϕ is the function in the integrand of (2), § 4.6, and a dot, as usual, denotes differentiation with respect to t, then

$$\int_{t_1}^{t_1'} K\left\{\alpha_1\frac{\partial(x,y)}{\partial(t,a_1)} + \alpha_2\frac{\partial(x,y)}{\partial(t,a_2)} + \mu\frac{\partial(x,y)}{\partial(t,\lambda)}\right\} dt = 0. \tag{6}$$

We now have three equations, two from (3), which holds for $t = t_1$ and $t = t_1'$, and equation (6). Eliminating α_1, α_2, μ we have

$$\begin{vmatrix} \dfrac{\partial(x,y)}{\partial(t_1,a_1)} & \dfrac{\partial(x,y)}{\partial(t_1,a_2)} & \dfrac{\partial(x,y)}{\partial(t_1,\lambda)} \\ \dfrac{\partial(x,y)}{\partial(t_1',a_1)} & \dfrac{\partial(x,y)}{\partial(t_1',a_2)} & \dfrac{\partial(x,y)}{\partial(t_1',\lambda)} \\ \displaystyle\int_{t_1}^{t_1'} K\frac{\partial(x,y)}{\partial(t,a_1)}\,dt & \displaystyle\int_{t_1}^{t_1'} K\frac{\partial(x,y)}{\partial(t,a_2)}\,dt & \displaystyle\int_{t_1}^{t_1'} K\frac{\partial(x,y)}{\partial(t,\lambda)}\,dt \end{vmatrix} = 0 \tag{7}$$

where $\dfrac{\partial(x,y)}{\partial(t_1,a_1)} = \dfrac{\partial(x,y)}{\partial(t,a_1)}$ with $t = t_1$.

The parameters of the end points of the range of integration of (5) and (6), § 4.6, are t_1 and t_2 and the Jacobi test then states that for a maximum or minimum value of I, as defined by equation (5), § 4.6, we must have $t_1 < t_2 < t_1'$ (or $t_1' < t_2 < t_1$).

For the Legendre test let

$$H(x,y,\dot{x},\dot{y}) = G(x,y,\dot{x},\dot{y})-\lambda\psi(x,y,\dot{x},\dot{y}), \tag{8}$$

where the functions G and ψ are those which occur in (5) and (6), § 4.6. Then it can be shown by differentiating (3) and (4), § 4.6, that

$$\frac{1}{\dot{y}^2}\frac{\partial^2 H}{\partial \dot{x}^2} = -\frac{1}{\dot{x}\dot{y}}\frac{\partial^2 H}{\partial \dot{x}\partial \dot{y}} = \frac{1}{\dot{x}^2}\frac{\partial^2 H}{\partial \dot{y}^2}. \tag{9}$$

The Legendre test then requires that these expressions maintain constant sign throughout the range of integration.

Denoting the expressions in (9) by P it can be shown by elaborate analysis that the second variation is equal to

$$\int\limits_{t_1}^{t_2} P\left(\dot{w}-\frac{\dot{q}}{q}w\right)^2 dt, \tag{10}$$

where q denotes the left-hand side of equation (3) above and w is a function of the variations $u_1(t)$ and $u_2(t)$, introduced in § 4.6.

With the help of this expression we can finally obtain the following result. If the characteristic equations (11) and (13), § 4.6, and the Jacobi and Legendre tests are all satisfied, then the integral I, of (1), § 4.6, is a maximum if the expressions in (9) are negative and a minimum if they are positive.†

To illustrate these results consider the problem of § 4.7, where it is required to find the plane curve of given length which encloses a maximum area. For this example:

$$H(x,y,\dot{x},\dot{y}) = \tfrac{1}{2}(x\dot{y}-\dot{x}y)-\lambda(\dot{x}^2+\dot{y}^2)^{\frac{1}{2}}, \tag{11}$$

and so

$$-\frac{1}{\dot{x}\dot{y}}\frac{\partial^2 H}{\partial \dot{x}\partial \dot{y}} = -\frac{\lambda}{(\dot{x}^2+\dot{y}^2)^{\frac{3}{2}}}. \tag{12}$$

Taking the positive value of the root it is then evident that the area enclosed is a maximum subject to possible limitations which may arise from the Jacobi test.

† For details of proofs of all statements in this section see Forsyth, loc. cit., pp. 60 and 387–406.

For the Jacobi test we shall use the solutions of the characteristic equations given by (9) and (10), § 4.7. Taking the parameter t to be equal to $\psi+\alpha$ and replacing a and b by a_1 and a_2 respectively, we have

$$x = a_1+\lambda\sin t, \quad y = a_2-\lambda\cos t. \tag{13}$$

The top row of the determinant in (7) evidently becomes

$$-\lambda\sin t_1, \quad \lambda\cos t_1, \quad -\lambda$$

and the second row can be obtained from these expressions by replacing t_1 by t_1'. For the third row we have $\phi = (\dot{x}^2+\dot{y}^2)^{\frac{1}{2}}$, so that the function K, defined by (5) above, is equal to $1/\lambda$, the curvature. The integrals in the third row of (7) are then easily evaluated, since λ is constant, giving us finally

$$\begin{vmatrix} -\lambda\sin t_1 & \lambda\cos t_1 & -\lambda \\ -\lambda\sin t_1' & \lambda\cos t_1' & -\lambda \\ \cos t_1'-\cos t_1 & \sin t_1'-\sin t_1 & -t_1'+t_1 \end{vmatrix} = 0.$$

This is an equation for t_1' and its root nearest to t_1 is found to be $t_1+2\pi$. Hence there is no conjugate point within a complete circumference and the circle solution of the problem gives an enclosed area which is a maximum without restriction, in agreement with the result obtained in § 4.10.

4.14. Subsidiary equations of non-integral type

Problems in which the equation of condition does not involve integration, as in § 4.8, are of such diverse nature that a general discussion of their second variation terms is beyond the scope of this book.

When non-holonomic restraints are absent the most important case is that of the geodesic, and then it is possible to reduce the problem to one dealt with in Chapter III. The reduction is effected by the use of Gaussian coordinates. The coordinates of a point P lying on a surface S can be expressed in terms of two parameters u and v. If u and v are made to depend upon a third parameter, t say, then as t varies the point P describes a curve lying wholly on S. The length of arc which joins two points whose parameters are t_1 and t_2 is then given by the integral I, where

$$I = \int_{t_1}^{t_2} (E\dot{u}^2 + 2F\dot{u}\dot{v} + G\dot{v}^2)^{\frac{1}{2}}\, dt \qquad (1)$$

and E, F, G are functions of u and v.†

The condition that the point should lie on the given surface is then implicit in the form of (1) and no further condition is necessary. The problem is thus reduced to that of finding the maximum and minimum of an integral whose integrand is a function of one independent variable, t, and of two parameters which in turn are also functions of t. Such problems have already been discussed in §§ 3.2 and 3.3, and they will be discussed again later in § 9.13 by means of a method introduced by Weierstrass.

EXAMPLE 10. Let (x_1, x_2, x_3, x_4) be the Cartesian coordinates of a point in four-dimensional space and let ds, the element of arc on a surface embedded in this space, be given by

$$ds^2 = \sum_{m=1}^{4} \sum_{n=1}^{4} g_{mn}\, dx_m\, dx_n,$$

where $g_{mn} = g_{nm}$. Show that the geodesics on the surface satisfy the equations

$$\sum_{m=1}^{4} g_{mr} \frac{d^2x_m}{ds^2} + \sum_{m=1}^{4} \sum_{n=1}^{4} [m, n; r] \frac{dx_m}{ds} \frac{dx_n}{ds} = 0 \quad (r = 1, 2, 3, 4),$$

where the Christoffel three-index symbol $[m, n; r]$ is defined by the equation

$$[m, n; r] = \frac{1}{2}\left(\frac{\partial g_{mr}}{\partial x_n} + \frac{\partial g_{nr}}{\partial x_m} - \frac{\partial g_{mn}}{\partial x_r}\right).$$

[Find the stationary value of the integral

$$\int \left(\sum_{m=1}^{4} \sum_{n=1}^{4} g_{mn}\, \dot{x}_m\, \dot{x}_n\right)^{\frac{1}{2}} dt$$

by means of the characteristic equations (9), § 3.2.]

† Courant, *Differential and Integral Calculus*, p. 102. See also § 9.13.

CHAPTER V

HAMILTON'S PRINCIPLE AND THE PRINCIPLE OF LEAST ACTION

5.1. Introduction

THE Calculus of Variations and the Principle of Least Action combine to form a powerful method of investigating problems of dynamics and mathematical physics, as the illustrations given in Chapters I and II show. Maupertuis, the author of the Principle of Least Action in 1744, declared it to be a metaphysical principle on which all the canons of motion are based. Professor E. B. Wilson† has said that Hamilton's Principle, which contains the Principle of Least Action as a special case, is the most fundamental and important single theorem in mathematical physics.

In this chapter we shall discuss in detail the theory of Hamilton's Principle and commence by defining three terms which are of frequent occurrence, all familiar to the student of dynamics.

5.2. Degrees of freedom

A particle in three-dimensional space requires three numbers to specify its position relative to a given frame or set of axes, for example its position can be specified by its three Cartesian coordinates (x, y, z). The particle is then said to have three degrees of freedom, one for each of its independent coordinates. If there are p particles all independent of each other, then we have a system with $3p$ independent coordinates, namely (x_i, y_i, z_i), $i = 1, 2,..., p$, and so there are $3p$ degrees of freedom. If constraints are imposed upon the particles, then the number of degrees of freedom is lowered. For example if the distance between the first two points is fixed and equal to l we have the relation

$$(x_1-x_2)^2+(y_1-y_2)^2+(z_1-z_2)^2 = l^2. \tag{1}$$

Hence if the values of y_1, z_1, x_2, y_2, z_2 are known, that of x_1 can be calculated. In this case the system has only $3p-1$ independent

† Professor of Mathematics at the Massachusetts Institute of Technology.

coordinates and it is then said to have $3p-1$ degrees of freedom.

It is evident that every constraint which is expressible in the form of one equation between some or all of the coordinates will lower the number of independent coordinates, and therefore the number of degrees of freedom, by one. Consequently a system consisting of p particles subjected to r such constraints has $3p-r$ degrees of freedom and requires $3p-r$ coordinates or parameters to specify its configuration at any instant of time.† In dynamics these parameters are generally denoted by q_i, $i = 1, 2,..., (3p-r)$, and their derivatives with respect to the time $\dot{q}_i$ are usually called the generalized velocities.

An unconstrained rigid body in three-dimensional space has six degrees of freedom since six coordinates are required to fix its position in relation to a given frame or set of axes. For example the three Cartesian coordinates of the centre of gravity and the three Eulerian‡ angles which measure the orientation of three fixed lines in the body in relation to a given frame fix the position of a rigid body in space.

In dynamics a rigid body is considered as if it were built up of a number of particles whose distances apart are kept invariable by suitable constraints. For purposes of visual imagination the particles can be considered as if they were situated at the vertices of a frame built of massless rigid rods smoothly jointed to each other at their extremities. Now each rod gives rise to an equation of restraint such as equation (1), so that if the frame consists of p particles and r rods, then the number of degrees of freedom is $3p-r$. But a rigid body in three dimensions has six degrees of freedom and so we must have $r = 3p-6$.

A little further consideration is required before we can leave the subject, for p particles can be joined by $\frac{1}{2}p(p-1)$ rods and, when $p > 4$, this quantity exceeds $3p-6$. It must therefore be noted that in order to attain rigidity by means of this imaginary frame *only* $3p-6$ rods are necessary. For example, if there are

† See Whittaker, *Analytical Dynamics*, p. 34, for an alternative definition of degrees of freedom.

‡ Whittaker, loc. cit., p. 9.

five particles numbered 1, 2, 3, 4, 5, then $\frac{1}{2}p(p-1) = 10$ and $3p-6 = 9$, so that nine rods are sufficient for rigidity. This is exhibited in Fig. V. 1, where the points 2, 3, 4 are joined by a triangle of light rigid rods and then particles 1 and 5 are each joined to 2, 3, and 4, the rods being smoothly jointed at their extremities. This is evidently sufficient to ensure rigidity without a connecting rod between particles 1 and 5.

If $3p-6$ rods are used to join the particles the resulting frame is called simply stiff and if more than $3p-6$ rods are used the frame is called overstiff.† If no external forces act on a simply stiff frame, then there are no internal stresses in the rods; and if a system of external forces acts on a simply stiff frame, then the stresses in the rods are determinate. This is not the case for overstiff frames which may be self stressed even if unacted on by any external forces. It is evident that the simply stiff frame which will turn a system of particles into a rigid body can be chosen in more than one way.

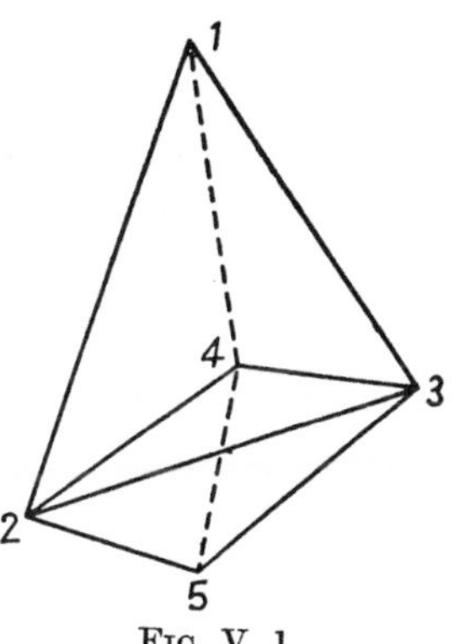

FIG. V. 1.

Since the stresses at the ends of each rod of the frame are equal and opposite it follows that for a rigid body conceived in this manner the internal forces are in equilibrium among themselves. This leads to D'Alembert's principle that the external forces and the reversed effective forces are in equilibrium during the motion of a rigid body.

All these statements are easily modified to suit the case of a plane rigid lamina moving in its own plane, the only essential difference from the results above being that for such a case we have three degrees of freedom instead of six. Hence if a plane rigid lamina is conceived as being built up of p coplanar particles, each having two degrees of freedom, the p particles must be joined by $2p-3$ massless rigid rods, smoothly jointed at their ends, in order to achieve rigidity.

† Routh, *Analytical Statics*, vol. i, § 150 et seq. Sometimes redundant or over-rigid may be used instead of overstiff.

5.3. Holonomic and non-holonomic systems

It sometimes happens that one or more of the particles are subjected to constraints of a different character from those discussed in § 5.2, for example, the equations of constraint may contain the velocities as well as the coordinates, or even time derivatives of higher order. In particular when one body rolls on a fixed body the point of contact of the moving body must be instantaneously at rest and the conditions for this to be the case must necessarily involve velocity terms. If those equations of constraint which involve time derivatives of the coordinates can be integrated so as to lead to equations involving the coordinates only (such as (1), § 5.2), then the system is said to be holonomic. If the equations of constraint cannot be so integrated, then the system is said to be non-holonomic.

This distinction can be put in another form. For a holonomic system the number of parameters required to specify the configuration of the system is equal to the number of degrees of freedom. But for a non-holonomic system the number of parameters required is greater than the number of degrees of freedom, the excess being equal to the number of non-integrable equations of constraint.

The distinction between the two systems is important since the equations of motion for the two cases differ in character.

To illustrate these remarks consider the following two cases. In the first case a vertical wheel, of radius a, rolls down the line of greatest slope of a perfectly rough non-horizontal plane inclined at an angle to the horizontal. After time t let x be the distance traversed by the centre and θ be the angle through which the wheel has turned. Then the condition that the point of contact should be instantaneously at rest is $\dot{x}-a\dot{\theta} = 0$. Since this equation can be integrated to $x-a\theta =$ constant, which does not involve the time, the system is holonomic.

For the second case consider the motion of a vertical wheel rolling on a perfectly rough horizontal plane so that it cannot slip sideways.† Let x and y be the coordinates of the point of

† This illustration is taken from Whittaker's *Analytical Dynamics*, p. 214.

contact and θ the angle between the x-axis and the plane of the wheel. Then, since the wheel cannot slip in a direction perpendicular to its plane, we must have

$$\dot{x}\sin\theta - \dot{y}\cos\theta = 0. \tag{1}$$

In general, θ is not constant. Therefore this equation is not integrable and the system is non-holonomic. This system has only two degrees of freedom, but its dynamical specification requires three parameters, x, y, and θ, together with the equation of constraint (1).

5.4. Conservative and non-conservative systems of force

Consider the work done by the forces of a field during the displacement of the system from one configuration C to another C_1. If the work is independent of the intermediate displacements between C and C_1 but depends only on the parameters required to specify C and C_1 the system is said to be conservative. Gravitational fields of force offer simple examples of conservative systems.

Conservative systems possess important properties. It is usual to fix the configuration C_1 permanently so that the work done is then a function of the parameters of the configuration C only. The work function is then usually called the potential function and denoted by V. If δs is an element of length, then $-\partial V/\partial s$ is equal to the force exerted by the field in the direction δs, and if $\delta\theta$ is an element of angular displacement about an axis (A), then $-\partial V/\partial\theta$ is the moment of the couple exerted by the system about the axis (A), in the same sense of rotation as $\delta\theta$. In vector language, for a conservative system the forces and moments of forces are expressible as the negative gradients of the scalar potential function of the system.†

Another important property associated with conservative systems of force is the Conservation of Energy (see § 5.14). This states the circumstances under which the sum of the kinetic and potential energies of a system is constant throughout the motion.

† Whittaker, loc. cit., p. 38.

5.5. Statement of Hamilton's principle

Let $q_1, q_2,..., q_n$ be the n independent parameters of a dynamical system and let us consider that the system is built up of a number of elementary particles joined by massless rigid rods as in § 5.2. Let m be the mass and (x, y, z) be the coordinates of a typical particle. Then the coordinates must be expressible in terms of the parameters q_i, so that we have expressions of the form

$$x = f(q_1, q_2,..., q_n; t) \tag{1}$$

together with corresponding relations for y and z.

The potential energy V is a function of the coordinates only and so must be expressible in the form

$$V = F(q_1, q_2,..., q_n). \tag{2}$$

If T denotes the kinetic energy, then

$$T = \sum_m \tfrac{1}{2}m(\dot{x}^2+\dot{y}^2+\dot{z}^2), \tag{3}$$

where $\sum_m$ denotes that the sum is taken over all the particles of the system. From (1) it follows that T is a function of the parameters of the form

$$T = \phi(q_1, q_2,..., q_n; \dot{q}_1, \dot{q}_2,..., \dot{q}_n; t), \tag{4}$$

where ϕ is of the second degree in $\dot{q}_1, \dot{q}_2,..., \dot{q}_n$.

The function $L = T-V$ is known as the Kinetic Potential.

Hamilton's principle then states that for a time interval with given end points t_0, t_1 the integral

$$\int_{t_0}^{t_1} L\, dt \tag{5}$$

is stationary when taken along an actual dynamical path. If, in addition, Jacobi's test is satisfied, i.e. the range of integration does not include any point conjugate to either of the extremities, then the integral (5) is a minimum.

In more detail let A, B, C, D denote the configurations through which the system passes during the time interval t_0 to t_1, A and D being the configurations at times t_0 and t_1 respectively, and B and C configurations for intermediate values of the time. Suppose now that in addition to the forces of the system other forces are imposed, by means of smooth constraints which do no work, so that the intermediate configurations are B_1 and C_1 instead of B and C. Then Hamilton's principle states that the value of the integral (5) taken along the actual dynamical path $ABCD$ is less than its value when taken along AB_1C_1D, subject of course to Jacobi's test being satisfied.

The principle can also be expressed in the following manner. If the differences between the two dynamical paths $ABCD$ and AB_1C_1D are expressible in terms of infinitesimals of the first order, then the differences between the corresponding values of integral (5) are of the second order in these infinitesimals.

When the forces are conservative, the principle as it stands is true both in the holonomic and non-holonomic cases. If the forces are not conservative, then the principle is still true, but in a modified form, as we shall see later in § 5.12.

The whole of ordinary dynamical theory, such as equations of motion, etc., can be deduced from Hamilton's principle.

5.6. Statement of the principle of least action

In the case of a conservative system of forces we have

$$T+V = k,$$

where k is constant. Hence $L = T-V = 2T-k$, so that

$$\int_{t_0}^{t_1} L\,dt = 2\int_{t_0}^{t_1} T\,dt-k(t_1-t_0). \tag{1}$$

Since t_1-t_0 is constant it follows that if Hamilton's principle is true then the integral

$$2\int_{t_0}^{t_1} T\,dt \tag{2}$$

is also stationary for a dynamical path and a minimum if Jacobi's test is satisfied. The integral (2) is known as the *action* and its stationary character is known in dynamics as *the principle of least action*.

For a single particle $2T = mv^2$ and $v\,dt = ds$, where s is the length of arc of the trajectory. The principle then becomes

$$m\int v\,ds, \tag{3}$$

when taken along the actual dynamical path, is a minimum.

This was the form in which the principle was stated by its author Maupertuis in 1744. His actual statement was as follows: 'L'Action est proportionnelle au produit de la masse par la vitesse et par l'espace. Maintenant, voici ce principe, si sage, si digne de l'Être suprême: Lorsqu'il arrive quelque changement dans la Nature, la quantité d'action employée pour ce changement est toujours la plus petite qu'il soit possible.'

The statement of the principle of least action by Thomson and Tait† is as follows: 'Of all the different sorts of paths along which a conservative system may be guided to move from one configuration to another, with the sum of its kinetic and potential energies constant, that one for which the action is least is such that the system will require only to be started with the proper velocities, to move along it unguided.'

5.7. Proof of Hamilton's principle: preliminary remarks

In the case when the system is conservative, the parameters being as usual $q_1, q_2,..., q_n$, the characteristic equations which ensure a stationary value for the integral

$$\int_{t_0}^{t_1} L\,dt \tag{1}$$

are given by theorem 7, § 3.2. They are

$$\frac{\partial L}{\partial q_i} - \frac{d}{dt}\left(\frac{\partial L}{\partial \dot{q}_i}\right) = 0 \tag{2}$$

$(i = 1, 2,..., n)$. Since V is a function of the coordinates only, we have $\partial V/\partial \dot{q}_i = 0$, so that equations (2) can be written in the form

$$\frac{d}{dt}\left(\frac{\partial T}{\partial \dot{q}_i}\right) - \frac{\partial T}{\partial q_i} = \frac{\partial V}{\partial q_i}. \tag{3}$$

These are the well-known Lagrangian equations of motion of a dynamical system. If they are proved dynamically the stationary property of integral (1) is an immediate consequence. This is the usual and probably one of the best methods of proving Hamilton's principle and will form the basis of our first proof. We shall also give another proof which is quite independent of

† *Natural Philosophy*, vol. i. 327. For this reference and Maupertuis's statement I am indebted to Forsyth, loc. cit., p. 363 footnote.

Lagrange's equations; it will be based upon the earlier and more fundamental notions of dynamics, namely Newton's laws of motion. The first variation of integral (1) will be dealt with in §§ 5.8 to 5.12 and the second variation in §§ 5.15 to 5.17.

5.8. First proof of Hamilton's principle for conservative holonomic systems

In this proof we first establish Lagrange's equations, (3), § 5.7, and so infer that (1), § 5.7, is a stationary integral for a dynamical path.

To establish these equations for a conservative holonomic system having n degrees of freedom, let a configuration be specified by the n parameters $q_1, q_2,..., q_n$ and let every rigid body be considered as built up of a number of elementary particles kept at invariable distances apart by means of a massless rigid simply stiff frame as described in § 5.2. Let m be the mass and (x, y, z) be the coordinates of a typical particle of the system, and let the external forces acting on this particle have components (X, Y, Z). By external forces we mean forces other than those in the rods of the simply stiff frames; the forces in these rods are called the internal forces. The components of the internal forces acting on m will be denoted by (X_i, Y_i, Z_i). Since the external forces form a conservative system, there exists a potential energy function V whose value is equal to the energy of the system.

From Newton's second law of motion we have

$$m\ddot{x} = X + X_i, \tag{1}$$

with corresponding equations for the y and z coordinates. But the internal forces of the frame must be in equal and opposite pairs and so are in equilibrium among themselves. Hence from (1) the system of forces, a typical one of which has components $[(X-m\ddot{x}), (Y-m\ddot{y}), (Z-m\ddot{z})]$, must also be in equilibrium. The principle of Virtual Work† then tells us that the work done by these forces in small displacements of the particles, namely

$$\sum_m \{(X-m\ddot{x})\,\delta x+(Y-m\ddot{y})\,\delta y+(Z-m\ddot{z})\,\delta z\}, \tag{2}$$

† Routh, *Analytical Statics*, vol. 1, chap. vi (2nd edition).

where $\sum_m$ denotes summation over all the particles of the system, is a quantity which is of the second order of smallness; δx, δy, δz being of the first order of smallness. Hence, if (2) is expressed in terms of the independent parameters $q_1, q_2,..., q_n$, the coefficients of $\delta q, \delta q,..., \delta q_n$ must each vanish.

Now x, y, z are functions of the parameters and of the time, t, given by expressions of the form

$$x = f(q_1, q_2,..., q_n; t), \tag{3}$$

with corresponding equations for y and z, and from this equation we can calculate δx in terms of δq_i $(i = 1, 2,..., n)$. In making the calculations it must be remembered that the displacements used in the principle of virtual work are purely geometrical and independent of the time, so that we have

$$\delta x = \frac{\partial x}{\partial q_1}\,\delta q_1 + \frac{\partial x}{\partial q_2}\,\delta q_2 + ... + \frac{\partial x}{\partial q_n}\,\delta q_n \tag{4}$$

with corresponding expressions for δy and δz. Inserting these values for δx, δy, δz in (2) and equating the coefficients of δq_i, $(i = 1, 2,..., n)$, to zero we have

$$\sum_m \left\{(X - m\ddot{x})\frac{\partial x}{\partial q_i} + (Y - m\ddot{y})\frac{\partial y}{\partial q_i} + (Z - m\ddot{z})\frac{\partial z}{\partial q_i}\right\} = 0 \tag{5}$$

$(i = 1, 2,..., n)$. These are essentially Lagrange's equations and the rest of the proof contains no further dynamical principles but consists of analytical transformations only. We shall transform these equations into the form (3), § 5.7, by means of two lemmas, one dealing with the potential energy terms and the other with the acceleration terms of (5)

LEMMA 1. $$\sum_m \left\{X\frac{\partial x}{\partial q_i} + Y\frac{\partial y}{\partial q_i} + Z\frac{\partial z}{\partial q_i}\right\} = -\frac{\partial V}{\partial q_i}. \tag{6}$$

PROOF. Vary the parameter q_i to $q_i + \delta q_i$ keeping all the other q's and t constant. From (4) it follows that the particle at (x, y, z) is displaced to

$$\left(x + \frac{\partial x}{\partial q_i}\,\delta q_i,\ y + \frac{\partial y}{\partial q_i}\,\delta q_i,\ z + \frac{\partial z}{\partial q_i}\,\delta q_i\right).$$

Hence the work lost by the system of external forces is equal to the left-hand side of (6) multiplied by δq_i.

But, from the definition of V in § 5.4, the energy gained by the system in the variation from q_i to $q_i+\delta q_i$ is $(\partial V/\partial q_i)\,\partial q_i$. Hence the right-hand side of (6) multiplied by δq_i is also equal to the work lost by the system and the truth of (6) is therefore established.

LEMMA 2. If x, y, z have continuous partial derivatives with respect to $q_1, q_2,..., q_n$; t up to at least the second order, then

$$m\ddot{x}\frac{\partial x}{\partial q_i} = \frac{d}{dt}\left\{\frac{\partial}{\partial \dot{q}_i}(\tfrac{1}{2}m\dot{x}^2)\right\} - \frac{\partial}{\partial q_i}(\tfrac{1}{2}m\dot{x}^2) \tag{7}$$

together with corresponding equations for y and z ($i = 1, 2,..., n$).

PROOF. From equation (3) we have

$$\dot{x} = \frac{\partial x}{\partial q_1}\dot{q}_1 + \frac{\partial x}{\partial q_2}\dot{q}_2 + ... + \frac{\partial x}{\partial q_n}\dot{q}_n + \frac{\partial x}{\partial t}. \tag{8}$$

It then follows that

$$\frac{\partial \dot{x}}{\partial \dot{q}_i} = \frac{\partial x}{\partial q_i} \tag{9}$$

and that

$$\frac{d}{dt}\left(\dot{x}\frac{\partial \dot{x}}{\partial \dot{q}_i}\right) - \dot{x}\frac{d}{dt}\left(\frac{\partial x}{\partial q_i}\right) = \ddot{x}\frac{\partial x}{\partial q_i}. \tag{10}$$

But

$$\frac{d}{dt}\left(\frac{\partial x}{\partial q_i}\right) = \frac{\partial^2 x}{\partial q_1 \partial q_i}\dot{q}_1 + \frac{\partial^2 x}{\partial q_2 \partial q_i}\dot{q}_2 + ... + \frac{\partial^2 x}{\partial q_n \partial q_i}\dot{q}_n + \frac{\partial^2 x}{\partial t \partial q_i}, \tag{11}$$

and since the partial derivatives are continuous we may write

$$\frac{\partial^2 x}{\partial q_i \partial q_m} = \frac{\partial^2 x}{\partial q_m \partial q_i}, \qquad \frac{\partial^2 x}{\partial t \partial q_i} = \frac{\partial^2 x}{\partial q_i \partial t}, \tag{12}$$

for all the relevant integral values of i and m. From (8), (11) then reduces to

$$\frac{d}{dt}\left(\frac{\partial x}{\partial q_i}\right) = \frac{\partial \dot{x}}{\partial q_i} \tag{13}$$

and so (10) can be written in the form

$$\frac{d}{dt}\left\{\frac{1}{2}\frac{\partial \dot{x}^2}{\partial \dot{q}_i}\right\} - \frac{1}{2}\frac{\partial \dot{x}^2}{\partial q_i} = \ddot{x}\frac{\partial x}{\partial q_i}. \tag{14}$$

Equation (7) then follows on multiplying by m. The corresponding equations for y and z are proved similarly.

We can now proceed with the proof of Lagrange's equations. Add (7) to the analogous equations for y and z, sum over all the particles, and let T denote the kinetic energy. Then

$$\sum_m m\left(\ddot{x}\frac{\partial x}{\partial q_i}+\ddot{y}\frac{\partial y}{\partial q_i}+\ddot{z}\frac{\partial z}{\partial q_i}\right) = \frac{d}{dt}\frac{\partial}{\partial \dot{q}_i}\Big\{\sum_m \tfrac{1}{2}m(\dot{x}^2+\dot{y}^2+\dot{z}^2)\Big\} - \\ -\frac{\partial}{\partial q_i}\Big\{\sum_m \tfrac{1}{2}m(\dot{x}^2+\dot{y}^2+\dot{z}^2)\Big\} \quad (15)$$

$$= \frac{d}{dt}\left\{\frac{\partial T}{\partial \dot{q}_1}\right\}-\frac{\partial T}{\partial q_i}. \quad (16)$$

Using (6) and (16) equation (5) then becomes

$$\frac{d}{dt}\left(\frac{\partial T}{\partial \dot{q}_i}\right)-\frac{\partial T}{\partial q_i} = -\frac{\partial V}{\partial q_i}. \quad (17)$$

The proof of Hamilton's principle now follows, since these equations are not only the equations of the dynamical paths but, by theorem 7, § 3.2, they are also the characteristic equations which ensure that $\int L\,dt$ is stationary.

5.9. Second proof of Hamilton's principle for conservative holonomic systems

The second proof commences by considering the motion of a single particle, of mass m, whose coordinates at time t are (x, y, z). This is acted on by a conservative force whose potential function V is a function of x, y, z only. We have

$$\int_{t_0}^{t_1} L\,dt = \int_{t_0}^{t_1} \{\tfrac{1}{2}m(\dot{x}^2+\dot{y}^2+\dot{z}^2)-V\}\,dt, \quad (1)$$

and from theorem 7, § 3.2, the integral is stationary if x, y, z satisfy

$$\frac{d}{dt}\left(\frac{\partial L}{\partial \dot{x}}\right)-\frac{\partial L}{\partial x} = 0, \quad (2)$$

together with two corresponding equations for y and z. Evaluating (2) and the corresponding equations we have

$$m\ddot{x} = -\frac{\partial V}{\partial x}, \quad m\ddot{y} = -\frac{\partial V}{\partial y}, \quad m\ddot{z} = -\frac{\partial V}{\partial z}, \quad (3)$$

which, according to Newton's second law, are the equations of motion of the particle. Hence for the dynamical path of a particle integral (1) is stationary.

The stationary character of (1) is unaltered by transferring to any other system of coordinates. For example, using spherical polar coordinates, it follows that for a dynamical path the integral $\int\limits_{t_0}^{t_1} \{\frac{1}{2}m(\dot{r}^2+r^2\dot{\theta}^2+r^2\dot{\psi}^2\sin^2\theta)-V\}\,dt$ must be stationary. From this may be deduced the equations of motion of a particle in spherical polar coordinates. The actual deduction is left as an exercise for the reader.

We now proceed to show that the principle is true for two particles. There are two cases to be considered. In the first case the two particles are independent of each other, as e.g. when two particles attract each other according to the Newtonian law of gravity. In the second case the particles are not independent of each other but are connected by a massless rigid rod so that their distance apart is invariable.

In the first case, in which there are six degrees of freedom, let m_1 and m_2 be the masses of the particles and (x_1, y_1, z_1), (x_2, y_2, z_2) be their respective coordinates. If V is the potential energy of the system, then

$$\int\limits_{t_0}^{t_1} L\,dt = \int\limits_{t_0}^{t_1} \{\tfrac{1}{2}m_1(\dot{x}_1^2+\dot{y}_1^2+\dot{z}_1^2)+\tfrac{1}{2}m_2(\dot{x}_2^2+\dot{y}_2^2+\dot{z}_2^2)-V\}\,dt. \quad (4)$$

Then, since the coordinates are all independent, according to theorem 7, § 3.2, there will be six characteristic equations to ensure that the integral (4) is stationary. They are

$$m\ddot{x}_1 = -\frac{\partial V}{\partial x_1}, \quad m\ddot{x}_2 = -\frac{\partial V}{\partial x_2}, \quad (5)$$

together with analogous equations for y_1, y_2, z_1, z_2. But these are once again the equations of motion of the particles according to Newton's second law of dynamics, so that (4) must be stationary for the dynamical paths.

It is evident that these arguments apply to any number of particles. In the case of n independent particles with $3n$ degrees of freedom in a conservative field of force the characteristic equations which ensure that $\int L\,dt$ should be stationary are also the equations of motion according to Newton's laws.

We now deal with the second case in which the two particles are connected by a massless rigid rod so as to maintain an invariable distance between them. Equation (4) above still holds, but in this case we have five degrees of freedom only and in addition to (4) we have

$$(x_2-x_1)^2+(y_2-y_1)^2+(z_2-z_1)^2 = l^2, \tag{6}$$

where l is the length of the rod.

The equations which ensure that $\int L\,dt$ should be stationary in this case can be obtained from the results of § 4.8. Theorem 13, § 4.8, which deals with integrals having two dependent and one independent variable together with one subsidiary conditional equation, is easily extended by the same arguments to the case of integrals with n dependent variables and m ($< n$) subsidiary conditional equations. The generalizations of (10) and (12), § 4.8, give us the characteristic equations which ensure that (4) should be stationary when subjected to the restriction (6).

They are

$$m_1\ddot{x}_1 = -\frac{\partial V}{\partial x_1}+\mu(x_2-x_1), \tag{7}$$

$$m_2\ddot{x}_2 = -\frac{\partial V}{\partial x_2}-\mu(x_2-x_1), \tag{8}$$

together with analogous equations for y_1, y_2, z_1, z_2, where μ is the same function for all six equations. These six equations together with equation (6) determine the seven quantities x_1, y_1, z_1, x_2, y_2, z_2, μ uniquely in terms of t, the time.

Now if F denotes the thrust in the connecting rod, then the component parallel to the x-axis acting on the first particle is $-F(x_2-x_1)/l$. On writing down the equations of motion of each particle, according to Newton's laws, we obtain six equations. It is easily seen that the two x equations are the same as (7) and (8), except that μ has been replaced by F/l, and that similar remarks apply to the corresponding y and z equations. But these together with (6) give x_1, y_1, z_1, x_2, y_2, z_2, F uniquely in terms of t. But we can pass from one set of equations to the other by a change of terminology in which μ is replaced by F/l. Therefore $\mu = F/l$ and the dynamical equations are the same as the characteristic equations (7), (8), etc. Hence $\int L\,dt$ is stationary for actual dynamical paths of the particles.

This stationary property is independent of the coordinates chosen and is still true if the system is reduced to one with five independent parameters, e.g. by eliminating one of the coordinates from L by means of equation (6).

We now deal with the general case of a rigid body. This, as in § 5.2, we envisage as built up of n particles situated at the vertices of a light rigid simply stiff frame of $3n-6$ rods, thus leaving the body with six degrees of freedom. Let m_i be the mass and (x_i, y_i, z_i) be the coordinates of a typical particle, which we shall refer to as particle i. Then

$$\int\limits_{t_0}^{t_1} L\,dt = \int\limits_{t_0}^{t_1} \left\{ \sum_{i=1}^{n} \tfrac{1}{2}m_i(\dot{x}_i^2+\dot{y}_i^2+\dot{z}_i^2)-V \right\} dt. \tag{9}$$

If the particles i and e are joined by a rod of length l_{ie} we have the equation

$$(x_i-x_e)^2+(y_i-y_e)^2+(z_i-z_e)^2 = l_{ie}^2, \tag{10}$$

which is one of $3n-6$ such equations, one for each rod of the frame. If particles p and q are not joined by a rod, then for them there is no equation such as (10), the invariance of their distance apart being a consequence of the $3n-6$ equations already in existence.

From the results of § 4.8, and in particular the generalization of theorem 13, § 4.8, the conditions that the integral (9) should be stationary are embodied in $3n$ equations of which

$$m\ddot{x}_i = -\frac{\partial V}{\partial x_i} + \sum_e \mu_{ie}(x_i-x_e) \tag{11}$$

is typical. Here $i = 1, 2,..., n$ and there are n analogous equations for y_i and for z_i $(i = 1, 2,..., n)$. μ_{ie} is an undetermined multiplier which may be a function of the coordinate and which satisfies the condition $\mu_{ei} = -\mu_{ie}$. The symbol $\sum\limits_e$ denotes that the sum is taken over all particles e which are joined to particle i by a rod of the frame.

We now proceed to count up the unknowns and the number of equations we have to determine them. There is a μ_{ie} for each equation such as (10), i.e. $3n-6$ of them, and there are $3n$ coordinates, making $6n-6$ unknowns in all. On the other hand, there are $3n-6$ equations such as (10) and $3n$ equations such as (11), so that we have sufficient to determine the coordinates and the undetermined multipliers uniquely in terms of the time, t.

Consider now the problem from the point of view of Newtonian dynamics and let F_{ie} denote the thrust in the rod joining particles i and e whenever such a rod exists. By means of Newton's second law write down the equation of motion of particle i parallel to the x-axis. A result is then obtained which is the same as (11) above except that μ_{ie} is replaced by F_{ie}/l_{ie}. Since the dynamical equations also determine the coordinates and the quantities F_{ie} uniquely in terms of t, it follows (i) that $\mu_{ie} = F_{ie}/l_{ie}$, and (ii) that the dynamical equations and the characteristic equations which ensure that (9) is stationary are the same. Hence for a dynamical path $\int L\,dt$ must be stationary.

At the same time we have also obtained an interesting relationship between the undetermined multipliers μ_{ie} and the internal forces.

This stationary property of $\int L\,dt$ is independent of the coordinates used and is still true if L is expressed in terms of six independent parameters, such as are required to give six degrees of freedom to a rigid body. The above proof although restricted to the case of one rigid body is easily extended to the case of several such bodies and such an extension completes the proof of Hamilton's principle for conservative fields of force.

5.10. First proof of Hamilton's principle for non-holonomic systems

Non-holonomic dynamical systems, defined in § 5.3, arise whenever there are non-integrable relations between the coordinates q_i and their time derivatives. We shall deal with relations of the type:

$$A_{1i}\dot{q}_1+A_{2i}\dot{q}_2+\ldots+A_{ni}\dot{q}_n = 0 \tag{1}$$

($i = 1, 2,\ldots, p$, $p < n$), where the A's are functions of the parameters q_i and t, the time. It is sometimes more convenient to use (1) in the equivalent form

$$A_{1i}\,\delta q_1 + A_{2i}\,\delta q_2 + \ldots + A_{ni}\,\delta q_n = 0. \tag{2}$$

In the proof which follows we shall not assume that (1) or (2) can be integrated.

On making small geometrical displacements consistent with the p equations (2), and noting that the internal forces do no work, it follows that the virtual work is given by (2), § 5.8. If the external forces are conservative and possess a potential function V, then lemmas 1 and 2, § 5.8, still remain true and we may write (2), § 5.8, in the form

$$\sum_{i=1}^{n} \left\{ -\frac{\partial V}{\partial q_i} - \frac{d}{dt}\left(\frac{\partial T}{\partial \dot{q}_i}\right) + \frac{\partial T}{\partial q_i} \right\} \delta q_i. \tag{3}$$

As in § 5.8, D'Alembert's principle tells us that (3) is the virtual work of a system of forces in equilibrium and so must be of the second order of small quantities, where δq_i $(i = 1, 2, \ldots, n)$ are all of the first order. But the q_i's are no longer independent parameters, as in § 5.8, since they are subjected to p constraints of the type (1). Hence we cannot equate each coefficient to zero, as in the holonomic case of § 5.8, but must proceed as follows.

Since there are p equations such as (2), then there are $n-p$ independent parameters. Multiply (2) by λ_i, so far an undetermined multiplier, and then subtract from the virtual work expression (3). Choose λ_i $(i = 1, 2, \ldots, n)$ so that the coefficients of δq_r $(r = 1, 2, \ldots, p)$ all vanish, giving us

$$\frac{d}{dt}\left(\frac{\partial T}{\partial \dot{q}_r}\right) - \frac{\partial T}{\partial q_r} = -\frac{\partial V}{\partial q_r} - \sum_{i=1}^{p} \lambda_i A_{ri} \tag{4}$$

$(r = 1, 2, \ldots, p)$. The expression for virtual work then becomes,

$$\sum_{r=p+1}^{n} \left\{ -\frac{\partial V}{\partial q_r} - \frac{d}{dt}\left(\frac{\partial T}{\partial \dot{q}_r}\right) + \frac{\partial T}{\partial q_r} - \sum_{i=1}^{p} \lambda_i A_{ri} \right\} \delta q_r. \tag{5}$$

Since the $n-p$ parameters $q_{p+1}, q_{p+2}, \ldots, q_n$ are independent of each other it follows that (5) can be a second-order quantity only if each coefficient of δq_r $(r = p+1, p+2, \ldots, n)$ vanishes.

On equating these coefficients to zero we obtain equations identical with (4) except that r has the values

$$r = p+1, p+2,..., n.$$

It is important to note that λ_i is independent of r.

Equations (4), now proved true for $r = 1, 2,..., n$, are the Lagrange equations for the non-holonomic case. Together with the p equations (1) they suffice to determine uniquely the n parameters q_i and the p undetermined multipliers λ_i. If we use the kinetic potential $L = T-V$ equations (4) become

$$\frac{\partial L}{\partial q_r}-\frac{d}{dt}\left(\frac{\partial L}{\partial \dot{q}_r}\right)-\sum_{i=1}^{p}\lambda_i A_{ri} = 0 \tag{6}$$

$$(r = 1, 2,..., n).$$

This completes the dynamical part of the investigation and we now proceed to find the characteristic equations which ensure that $\int L\,dt$ is stationary when subjected to the restraints given by equations (1).

From (12), § 4.11, these characteristic equations are

$$\frac{\partial L}{\partial q_r}-\frac{d}{dt}\left(\frac{\partial L}{\partial \dot{q}_r}\right)-\sum_{i=1}^{p}\nu_i A_{ri} = 0 \tag{7}$$

$(r = 1, 2,..., n)$, where ν_i is independent of r. Once again these equations together with equations (1) serve to determine the r parameters q_i and the p multipliers ν_i uniquely. By comparison of (6) and (7) we conclude that $\lambda_i = \nu_i$ $(i = 1, 2,..., p)$ and that the dynamical and characteristic equations are the same. Hence $\int L\,dt$ is stationary for a dynamical path.

5.11. Second proof of Hamilton's principle for non-holonomic systems

We now proceed to extend the ideas used in § 5.9 to the non-holonomic case. A non-holonomic system can always be reduced to a holonomic one by the introduction of extra forces which (i) do no work and (ii) guide the system so that the equations of constraint are satisfied.

Consider the illustration given in § 5.3 of a vertical wheel rolling on a perfectly rough horizontal plane without slipping sideways. Let (x, y) be the coordinates of the point of contact and θ the angle between the x-axis and the plane of the wheel. Then for no side slipping we must have

$$\dot{x}\sin\theta - \dot{y}\cos\theta = 0. \tag{1}$$

This non-integrable equation shows that the system is non-holonomic.

Now apply to the point of contact forces X and Y, parallel to the x and y axes respectively, so that no work is done during the motion. This requires that

$$X\,\delta x + Y\,\delta y = 0. \tag{2}$$

By comparison of (1) and (2) it follows that

$$X = \lambda\sin\theta, \qquad Y = -\lambda\cos\theta, \tag{3}$$

where λ is a convenient factor. Thus by introducing these forces the equation of constraint (1) can be dispensed with and the system reduced to a holonomic one.

In the general case, considering that a rigid body is made up of n particles situated at the vertices of a rigid massless frame as in § 5.2, let a typical particle be of mass m_i and have coordinates (x_i, y_i, z_i). Let the external forces acting on particle i have components (X_i, Y_i, Z_i) parallel to the axes and suppose that the system is rendered holonomic by the introduction of additional forces acting on particle i whose components are (P_i, Q_i, R_i).

Since these additional forces do no work we must have the equation†

$$\sum_{i=1}^{n}(P_i\,\delta x_i + Q_i\,\delta y_i + R_i\,\delta z_i) = 0. \tag{4}$$

Also, if V is the potential of the external forces we have

$$\int_{t_0}^{t_1} L\,dt = \int_{t_0}^{t_1}\left\{\sum_{i=1}^{n}\tfrac{1}{2}m_i(\dot{x}_i^2+\dot{y}_i^2+\dot{z}_i^2) - V\right\}dt \tag{5}$$

together with

$$(x_i-x_e)^2+(y_i-y_e)^2+(z_i-z_e)^2 = l_{ie}^2, \tag{6}$$

where l_{ie} is the length of the rod which joins particles i and e. There are $(3n-6)$ equations such as (6).

† If there are q equations such as (4) then there will be q constants such as σ in (7). But the arguments are essentially unaltered.

The characteristic equations which ensure a stationary value for $\int L\,dt$, when subjected to constraints (4) and (6), are given by (12), § 4.11. There are $3n$ such equations, one for each coordinate of the n particles, a typical one being

$$-\frac{\partial V}{\partial x_i}-m_i\ddot{x}_i-\sum_e \mu_{ie}(x_i-x_e)-\sigma P_i = 0. \tag{7}$$

Here $\sum_e$ denotes a sum over all the particles e which are connected to particle i by a rod of the frame which gives rigidity to the system and σ is an undetermined multiplier. The total number of equations (4), (6), and (7) is $6n-5$, and there are thus sufficient to determine the values of the $3n$ coordinates of the n particles, the values of the $3n-6$ multipliers μ_{ie}, and also that of σ in terms of t, the time.

Now consider the equations of motion of the particles according to Newton's laws. If F_{ie} denotes the thrust in the rod joining particles i and e, then the equation of motion parallel to the x-axis of particle i is the same as (7) except that μ_{ie} is replaced by $-F_{ie}/l_{ie}$ and σ by -1. Since the dynamical equations determine the coordinates and F_i uniquely in terms of t, it follows that $\mu_{ie} = -F_{ie}/l_{ie}$ and $\sigma = -1$. Hence once more the dynamical and the characteristic equations are the same and so for a dynamical path $\int L\,dt$ is stationary.

This completes the proof of Hamilton's principle when the system is non-holonomic.

5.12. Hamilton's principle for non-conservative dynamical systems

If we except lemma 1 the analysis of § 5.8 is independent of the nature of the external forces and is therefore still valid when they are not conservative. Using the terminology of § 5.8, let δW denote the work done by the external forces in a small displacement so that

$$\delta W = \sum_m (X\,\delta x+Y\,\delta y+Z\,\delta z). \tag{1}$$

For a conservative system of forces $W = -V$ but for a non-conservative system W depends upon the path of integration and is no longer a function of position.

From (4), § 5.8 we have

$$\delta W = \sum_{i=1}^{n} Q_i \delta q_i, \tag{2}$$

where
$$Q_i = \sum_m \left\{X \frac{\partial x}{\partial q_i} + Y \frac{\partial y}{\partial q_i} + Z \frac{\partial z}{\partial q_i}\right\}. \tag{3}$$

$Q_i \delta q_i$ is the work done by the external forces in a displacement where all the parameters except q_i are kept constant and q_i is increased to $q_i + \delta q_i$.

If we replace (6), § 5.8, by (3) above and use (5), § 5.8 and lemma 2, § 5.8, we deduce that

$$\frac{d}{dt}\left(\frac{\partial T}{\partial \dot{q}_i}\right) - \frac{\partial T}{\partial q_i} = Q_i. \tag{4}$$

This is the form which the Lagrange equations take when the external forces are not conservative.

In the case of conservative external forces we may write the Hamilton's principle in the following manner: let

$$I = \int_{t_0}^{t_1} (T - V)\, dt = \int_{t_0}^{t_1} (T + W)\, dt, \tag{5}$$

then $\delta I = 0$ along a dynamical path. In the case of non-conservative forces the statement of Hamilton's principle is that

$$\delta \int_{t_0}^{t_1} T\, dt + \int_{t_0}^{t_1} \sum_{i=1}^{n} Q_i \delta q_i\, dt = 0 \tag{6}$$

along a dynamical path. For evidently from (2) the variation (6) yields characteristic equations which are identical with (4).

5.13. Proof of Lagrange's equations of motion

The proofs of Hamilton's principle given in §§ 5.9 and 5.11 depend upon Newton's laws of motion and make no appeal to Lagrange's equations. These equations may then be legitimately deduced from the results of §§ 5.9 and 5.11.

The Lagrangian equations are an immediate consequence of Hamilton's principle, for on expressing L, the Kinetic Potential in terms of the parameters $q_1, q_2,..., q_n$, the characteristic equations which ensure that $\int_{t_0}^{t_1} L\,dt$ is stationary are the Lagrange equations. In the holonomic case they are in the form given by (2) (or (3)), § 5.7. In the non-holonomic case, where we have non-integrable equations of the type

$$A_{1,i}\dot{q}_1+A_{2,i}\dot{q}_2+...+A_{n,i}\dot{q}_n = 0, \tag{1}$$

they are in the form given by (6), § 5.10.

5.14. The energy equation for conservative fields of force

For conservative fields of force there is frequently an important first integral of the equations of motion known as the energy equation.

If the time is not explicit in the equations relating the coordinates of a point to the parameters, then (3), § 5.8, becomes

$$x_i = f_i(q_1, q_2,..., q_n), \tag{1}$$

with corresponding equations for y and z. The kinetic energy T, given by

$$T = \sum_m \tfrac{1}{2}m(\dot{x}^2+\dot{y}^2+\dot{z}^2), \tag{2}$$

is then a function of q_i and $\dot{q}_i$ $(i = 1, 2,..., n)$, which is homogeneous and of degree two in $\dot{q}_i$ $(i = 1, 2,..., n)$. Hence, by Euler's theorem on homogeneous functions, we have

$$\sum_{i=1}^{n} \dot{q}_i \frac{\partial T}{\partial \dot{q}_i} = 2T. \tag{3}$$

Lagrange's equations in the holonomic case state that

$$\frac{d}{dt}\left(\frac{\partial T}{\partial \dot{q}_i}\right)-\frac{\partial T}{\partial q_i} = -\frac{\partial V}{\partial q_i}; \tag{4}$$

and on multiplying by $\dot{q}_i$ and rearranging we have

$$\frac{d}{dt}\left(\dot{q}_i \frac{\partial T}{\partial \dot{q}_i}\right)-\ddot{q}_i \frac{\partial T}{\partial \dot{q}_i}-\dot{q}_i \frac{\partial T}{\partial q_i} = -\dot{q}_i \frac{\partial V}{\partial q_i}. \tag{5}$$

If we sum this result for all values of $i = 1, 2,..., n$ and use (3) we get

$$\frac{d}{dt}(2T)-\frac{dT}{dt} = -\frac{dV}{dt}, \tag{6}$$

which integrates immediately to

$$T+V = \text{constant}, \tag{7}$$

the desired energy equation.

For the non-holonomic case, with constraints of the type

$$A_{1m}\dot{q}_1+A_{2m}\dot{q}_2+\ldots+A_{nm}\dot{q}_n = 0 \tag{8}$$

$(m = 1, 2,\ldots, p < n)$, the Lagrange equations take the form

$$\frac{d}{dt}\left(\frac{\partial T}{\partial \dot{q}_i}\right)-\frac{\partial T}{\partial q_i} = -\frac{\partial V}{\partial q_i}-\sum_{m=1}^{p}\lambda_m A_{im}, \tag{9}$$

as proved in § 5.10. Multiplying this equation by $\dot{q}_i$ and summing for $i = 1, 2,\ldots, n$ the terms involving A_{im} cancel out, by virtue of (8), so that (6) and (7) still remain true for the non-holonomic case.

Equation (7) is also frequently true for non-dynamical problems in which the characteristic equations are of the Lagrange type.

5.15. The second variation

The study of the second variation of $\int L\,dt$ is necessarily difficult. The account given here depends upon two facts, the first that a small variation in the characteristic curves effect a change in $\int L\,dt$ which is largely dependent upon the kinetic energy, and the second that the kinetic energy must be positive or zero. A function whose value is always positive is called positive definite and one whose value is always either positive or zero is called positive semi-definite.† The kinetic energy is evidently included in the second category.

In the proof we shall confine ourselves to the case where there is conservation of energy in the form given by (7), § 5.14, and consider only the special variation in the characteristic curves defined in the next section. It can be shown that the general weak variation of the characteristic curves can be expressed as the sum of special variations of the type considered here, but

† A function whose value is always negative is called negative definite and one whose value is always either negative or zero is called negative semi-definite.

for the general case the reader is referred to *Variationsrechnung* by C. Carathéodory.†

As pointed out in § 5.14, the kinetic energy is a quadratic function of $\dot{q}_1$ and we may write

$$2T = \sum_{i=1}^{n} \sum_{m=1}^{n} a_{im}\dot{q}_i\dot{q}_m, \tag{1}$$

where a_{im} is a function of the parameters q_i ($i = 1, 2,..., n$), but not of $\dot{q}_i$ ($i = 1, 2,..., n$). We may assume, without loss of generality, that $a_{im} = a_{mi}$.

The fact that T is a positive semi-definite function is a property of coefficients a_{im}, as we shall prove in § 5.17; in other words, because (1) is positive semi-definite then all expressions of the type

$$\sum_{i=1}^{n} \sum_{m=1}^{n} a_{im} b_i b_m \tag{2}$$

are also positive semi-definite for all values of b_i ($i = 1, 2,..., n$).

5.16. A special variation of the extremals

The characteristic curves, or dynamical orbits, for which $\int L\,dt$ is stationary are given by Lagrange's equations (3), § 5.7. Consider the deviation from these curves tabulated below. Here $q_1, q_2,..., q_n$ are the values of the parameters for those dynamical paths which make $\int L\,dt$ stationary and $\bar{q}_1; \bar{q}_2,..., \bar{q}_n$ denote the values for the deviated path (see Fig. V. 2).

Interval of time	*Value of* $\bar{q}_i$ ($i = 1, 2,..., n$)	*Consequent value of* $\dot{\bar{q}}_i$
t_0 to $t'-h$	$\bar{q}_i = q_i$	$\dot{\bar{q}}_i = \dot{q}_i$
$t'-h$ to t'	$\bar{q}_i = q_i+(t-t'+h)\alpha_i$	$\dot{\bar{q}}_i = \dot{q}_i+\alpha_i$
t' to $t'+h$	$\bar{q}_i = q_i+(h+t'-t)\alpha_i$	$\dot{\bar{q}}_i = \dot{q}_i-\alpha_i$
$t'+h$ to t_i	$\bar{q}_i = q_i$	$\dot{\bar{q}}_i = \dot{q}_i$

In this table h is a positive constant and α_i ($i = 1, 2,..., n$), are convenient constants, all independent of the time t. Also $t_0 < t'-h < t'+h < t_i$.

Denote the kinetic potential in the two cases by $L(q_i, \dot{q}_i, t)$ and $L(\bar{q}_i, \dot{\bar{q}}_i, t)$ and the corresponding values of $\int L\,dt$ by I and $\bar{I}$.

† B. G. Teubner (Leipzig), chap. xii. For an alternative discussion see Forsyth, loc. cit., p. 371 et seq.

Since q_i is varied inside the time-interval $t'-h$ to $t'+h$ only, we have

$$\bar{I}-I = \int\limits_{t'-h}^{t'+h} \{L(\bar{q}_i, \dot{\bar{q}}_i, t)-L(q_i, \dot{q}_i, t)\}\, dt \tag{1}$$

$$\left.\begin{aligned} &= \int\limits_{t'-h}^{t'+h} \{L(\bar{q}_i, \dot{\bar{q}}_i, t)-L(q_i, \dot{\bar{q}}_i, t)\}\, dt \\ &+ \int\limits_{t'-h}^{t'+h} \{L(q_i, \dot{\bar{q}}_i, t)-L(q_i, \dot{q}_i, t)\}\, dt. \end{aligned}\right\} \tag{2}$$

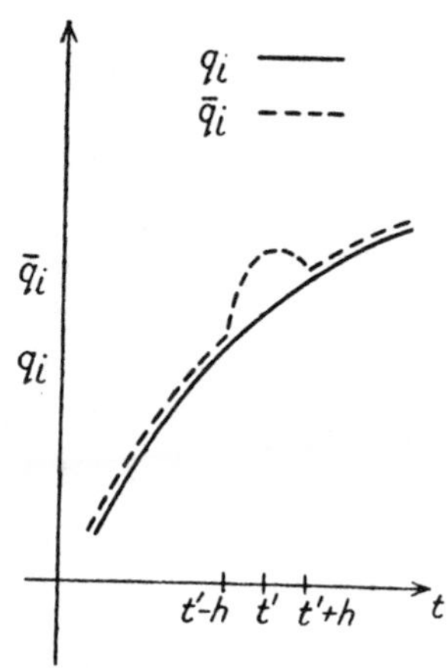

FIG. V. 2.

We shall take h to be small and make use of the Landau O notation defined in § 1.2. We can now prove with little calculation that the first integral of (2) is $O(h^2)$. For, by the first mean-value theorem, the integrand of the first integral is the sum of n terms of the type

$$(\bar{q}_i-q_i)\frac{\partial}{\partial q_i} L(\bar{\bar{q}}_i, \dot{\bar{q}}_i, t), \tag{3}$$

where $q_i < \bar{\bar{q}}_i < \bar{q}_i$. From the table we have

$$\int\limits_{t'-h}^{t'+h} (\bar{q}_i-q_i)\, dt = \int\limits_{t'-h}^{t'} (t-t'+h)\alpha_i\, dt + \int\limits_{t'}^{t'+h} (h+t'-t)\alpha_i\, dt = \alpha_i h^2. \tag{4}$$

Now let M_i be the greatest value of $\left|\frac{\partial}{\partial q_i} L(\bar{\bar{q}}_i, \dot{\bar{q}}_i, t)\right|$ in the interval $t'-h$ to $t'+h$, and let M be the greatest of the quantities $M_1, M_2,..., M_n$. Then since $\bar{q}_i - q_i$ has the same sign throughout the interval $t'-h$ to $t'+h$ it follows that

$$\left|\int_{t'-h}^{t'+h} \{L(\bar{q}_i, \dot{\bar{q}}_i, t) - L(q_i, \dot{\bar{q}}_i, t)\}\, dt\right| \leqslant nMh^2 \tag{5}$$

$$= O(h^2). \tag{6}$$

In the second integral of (2) only changes in $\dot{q}_i$ are encountered. Hence the value of the integrand is due entirely to changes in the kinetic energy and is independent of the potential energy. But a_{im} is a function of q_i only, so that from (1) we have

$$L(q_i, \dot{\bar{q}}_i, t) - L(q_i, \dot{q}_i, t) = \tfrac{1}{2}\sum_{i=1}^{n}\sum_{m=1}^{n} a_{im}(\dot{\bar{q}}_i \dot{\bar{q}}_m - \dot{q}_i \dot{q}_m) \tag{7}$$

$$= \sum_{i=1}^{n} (\dot{\bar{q}}_i - \dot{q}_i)\frac{\partial L}{\partial \dot{q}_i} + \tfrac{1}{2}\sum_{i=1}^{n}\sum_{m=1}^{n} a_{im}\alpha_i \alpha_m, \tag{8}$$

where $L = L(q_i, \dot{q}_i, t)$. We may deduce (8) from (7) either by using the table or by expanding the left-hand side of (7) in powers of $(\dot{\bar{q}}_i - \dot{q}_i)$ $(i = 1, 2,..., n)$, with the help of Taylor's theorem.

On integrating by parts and noting that $(\bar{q}_i - q_i)$ vanishes at both limits of integration we have

$$\int_{t'-h}^{t'+h} (\dot{\bar{q}}_i - \dot{q}_i)\frac{\partial L}{\partial \dot{q}_i}\, dt = \int_{t'-h}^{t'+h} \frac{d}{dt}\left(\frac{\partial L}{\partial \dot{q}_i}\right)(\bar{q}_i - q_i)\, dt \tag{9}$$

$$= \int_{t'-h}^{t'+h} \frac{\partial L}{\partial q_i}(\bar{q}_i - q_i)\, dt \tag{10}$$

on using Lagrange's equations which are true for the characteristic curves. But the integrand of (10) has the same form as (3), so that by the arguments applied to (3) we have

$$\int_{t'-h}^{t'+h} \sum_{i=1}^{n} (\dot{\bar{q}}_i - \dot{q}_i)\frac{\partial L}{\partial q_i}\, dt = O(h^2). \tag{11}$$

Finally we must consider the integral

$$\int_{t'-h}^{t'+h} \tfrac{1}{2}\sum_{i=1}^{n}\sum_{m=1}^{n} a_{im}\alpha_i \alpha_m\, dt, \tag{12}$$

where a_{im} is a function of q_i (but not of $\dot{q}_i$), and is therefore a function of t. Denote the integrand of (12) by $E(t)$, then

$$\int_{t'-h}^{t'+h} \tfrac{1}{2}\sum_{i=1}^{n}\sum_{m=1}^{n} a_{im}\,\alpha_i\,\alpha_m\,dt = \int_{t'-h}^{t'+h} E(t')\,dt + \int_{t'-h}^{t'+h} \{E(t)-E(t')\}\,dt. \tag{13}$$

Now in the interval of variation of q_i we have, from the table above, $t-t' = O(h)$ and the range of integration is equal to $2h$. Hence the second integral on the right of (13) is $O(h^2)$, and we have

$$\int_{t'-h}^{t'+h} \tfrac{1}{2}\sum_{i=1}^{n}\sum_{m=1}^{n} a_{im}\,\alpha_i\,\alpha_m\,dt = 2hE(t')+O(h^2). \tag{14}$$

On assembling the results of equations (6), (8), (11), and (14) it follows that

$$\bar{I}-I = 2hE(t')+O(h^2). \tag{15}$$

This is a fundamental result. For sufficiently small positive h, it is evident that $\bar{I} \geqslant I$ if and only if $E(t') \geqslant 0$. Now let the interval t_0 to t_1 be subdivided into elements of length $2h$ and let variations such as those defined by the table above be imposed on the parameters q_i. If $\bar{I}$ denotes the consequent value of $\int L\,dt$, then $\bar{I} \geqslant I$ if and only if $E(t')$ is positive definite for all values of t' inside the interval t_0 to t_1. It follows that I is a minimum for the range of integration (t_0, t_1) if and only if $E(t) = \frac{1}{2}\sum_{i=1}^{n}\sum_{m=1}^{n} a_{im}\,\alpha_i\,\alpha_m$ is positive definite at all points of the range.

In § 5.18 we shall prove that $E(t)$ is positive semi-definite because the kinetic energy is positive semi-definite. If we anticipate this result we see that $\int L\,dt$, when taken along an extremal, is a minimum for a range which excludes points at which $E(t)$ vanishes. Points for which $E(t) = 0$ then remain to be dealt with.

5.17. Conjugate points

It is evident from (15), § 5.16, that if $E(t)$ vanishes at some point of the range (t_0, t_1), then a varied path can be found such that $\bar{I}-I = O(h^2)$, whereas if $E(t)$ does not vanish then this difference is $O(h)$.

We now show, by means of a contradiction, that if

$$\bar{I} - I = O(h^2)$$

then the varied path must also be an extremal for $\int L\,dt$. We shall once again anticipate the result of § 5.18 so that along an extremal $\int L\,dt$ is a minimum.

Let C denote the extremal, $\bar{C}$ the varied curve, and P_0, P_1 the points corresponding to t_0, t_1 respectively. If $\bar{C}$ is not an extremal we may take two points Q_0 and Q_1 on it, lying between P_0 and P_1, and draw the extremal through Q_0, Q_1, say $Q_0\,RQ_1$. Then the value of $\int L\,dt$ taken along $P_0\,Q_0\,RQ_1\,P_1$ must be less than when taken along $\bar{C}$. Hence, if these values are denoted by $\bar{\bar{I}}$ and $\bar{I}$, we have $\bar{\bar{I}} < \bar{I}$ and by choosing Q_0 and Q_1 suitably we can make $\bar{\bar{I}} = \bar{I} - |O(h)|$. Thus $\bar{\bar{I}} = I - O(h) + O(h^2)$ and so $\bar{\bar{I}} < I$, which is impossible if C is an extremal.

If $\bar{I} = I + O(h^2)$ and $\bar{C}$ is not a characteristic curve we are led to a contradiction. Therefore, from (15), § 5.16, it follows that $E(t)$ must vanish at points of intersection of neighbouring characteristic curves. If A is one of these points of intersection then, as shown in § 2.9, the others are the conjugate points or kinetic foci of A.

It is now evident that if the range of integration does not contain two points which are conjugates of each other then $E(t)$ can never vanish and Hamilton's principle may be stated as follows. For ranges of integration which exclude points conjugate to either end point the value of $\int L\,dt$, when taken along actual dynamical paths, is a minimum for weak variations.

In § 9.3 below it will be proved that $\int L\,dt$ admits a strong minimum for the dynamical path of a particle.

5.18. Positive semi-definite quadratic forms

To complete the theory the assumption made at the end of § 5.16, namely that $E(t)$ is positive semi-definite, remains to be justified.

Let
$$E = \sum_{i=1}^{n} \sum_{m=1}^{n} a_{im}\,\alpha_i\,\alpha_m, \tag{1}$$

where $a_{im} = a_{mi}$. Using the language of geometry, consider the quantities α_i $(i = 1, 2,..., n)$ as the coordinates of a point in n-dimensional space and the equation

$$\alpha_2^2+\alpha_2^2+...+\alpha_n^2 = 1 \tag{2}$$

as the equation of a hypersphere with unit radius.

It is easy to show that if $E \geqslant 0$ for all points on sphere (2) then $E > 0$ for all points of the space.

For if
$$\alpha_1^2+\alpha_2^2+...+\alpha_n^2 = k^2, \tag{3}$$

then α_i/k $(i = 1, 2,..., n)$ is a point on (2) so that by hypothesis

$$\sum_{i=1}^{n}\sum_{m=1}^{n} a_{im}\frac{\alpha_i\,\alpha_m}{k^2} \geqslant 0.$$

Consequently

$$\sum_{i=1}^{n}\sum_{m=1}^{n} a_{im}\,\alpha_i\,\alpha_m = k^2\sum_{i=1}^{n}\sum_{m=1}^{n} a_{im}\frac{\alpha_i\,\alpha_m}{k^2} \tag{4}$$

$$\geqslant 0. \tag{5}$$

The problem is then reduced to that of finding the maximum and minimum values of E subject to condition (2). By the usual theory of maxima and minima† the stationary values of E occur when

$$\frac{\partial E/\partial\alpha_1}{\alpha_1} = \frac{\partial E/\partial\alpha_2}{\alpha_2} = ... = \frac{\partial E/\partial\alpha_n}{\alpha_n} = \lambda, \tag{6}$$

where λ is an undetermined multiplier. This gives us

$$\frac{\sum_{i=1}^{n} a_{1i}\,\alpha_i}{\alpha_1} = \frac{\sum_{i=1}^{n} a_{2i}\,\alpha_i}{\alpha_2} = ... = \frac{\sum_{i=1}^{n} a_{ni}\,\alpha_i}{\alpha_n} = \lambda, \tag{7}$$

from which two deductions are easily made.

The first deduction is made by eliminating α_i $(i = 1, 2,..., n)$ from equations (7). The following equation is then obtained for λ,

$$\begin{vmatrix} a_{11}-\lambda & a_{12} & \cdot & \cdot & a_{1n} \\ a_{21} & a_{22}-\lambda & \cdot & \cdot & a_{2n} \\ \cdot & \cdot & \cdot & \cdot & \cdot \\ \cdot & \cdot & \cdot & \cdot & \cdot \\ a_{n1} & a_{n2} & \cdot & \cdot & a_{nn}-\lambda \end{vmatrix} = 0. \tag{8}$$

† Courant, *Differential and Integral Calculus*, vol. ii, p. 188.

The second deduction is obtained by multiplying the numerator and denominator of the first fraction of (7) by α_1, the second fraction similarly by α_2, etc., and then summing all the numerators and denominators. The result, on using (2), is

$$\lambda = \sum_{i=1}^{n} \sum_{m=1}^{n} a_{im} \alpha_i \alpha_m. \tag{9}$$

Hence λ is equal to one of the stationary values of E.

From these two deductions it follows that the roots of (8), considered as an equation in λ, are the stationary values of E.

Before proceeding further with the proof a digression is made in order to prove that if a_{im} is real then all the roots of (8) are real. For if λ is a complex root and $\bar{\lambda}$ is its conjugate then, by virtue of (8), there exist real or complex numbers $\mu_1, \mu_2, \ldots, \mu_n$, such that

$$\sum_{m=1}^{n} a_{im} \mu_m = \lambda \mu_i \tag{10}$$

$(i = 1, 2, \ldots, n)$.

If $\bar{\mu}_m$ is the conjugate complex of μ_m then, since a_{im} is real, we must also have

$$\sum_{m=1}^{n} a_{im} \bar{\mu}_m = \bar{\lambda} \bar{\mu}_i \tag{11}$$

$(i = 1, 2, \ldots, n)$.

Multiply (10) by $\bar{\mu}_i$ and (11) by μ_i and sum for $i = 1, 2, \ldots, n$ in each case. On subtracting one sum from the other and using $a_{im} = a_{mi}$, the left-hand sides cancel, leaving

$$(\lambda - \bar{\lambda}) \sum_{i=1}^{n} \mu_i \bar{\mu}_i = 0. \tag{12}$$

Since $\sum_{i=1}^{n} \mu_i \bar{\mu}_i$ cannot vanish it follows that $\lambda = \bar{\lambda}$ and therefore that λ must be real.

Reverting to the main proof, since the stationary values of E on the hypersphere (2) are equal to the roots of equation (8) it follows that if all the roots of (8) are positive, then all the maximum and minimum values of E are positive, and so E is positive definite. Similarly if all the roots are either positive or zero, then E is positive semi-definite. It is also obvious that if all the roots of (8) are negative then E is negative definite, and if the roots are negative or zero then E is negative semi-definite.

Now the roots of (8) depend for their values entirely upon the quantities a_{im} and are independent of the quantities α_i. Hence the property of being positive definite (or positive semi-definite) must depend only upon the values of the a_{im} and not upon the values of α_i. But T, the kinetic energy, must be positive or zero since it is of the form $\sum_m \frac{1}{2}m(\dot{x}^2+\dot{y}^2+\dot{z}^2)$. Also it remains positive or zero when transformed to generalized coordinates. Hence $\sum\sum a_{im}\dot{q}_i\dot{q}_m$, which from (1), § 5.15 is equal to $2T$, is positive semi-definite and consequently $\sum_{i=1}^{n}\sum_{m=1}^{n} a_{im}\alpha_i\alpha_m$ must also be positive semi-definite.

This proves an assumption made in the last part of § 5.16 and so completes the proof of Hamilton's principle.

5.19. A particle under no forces describes a geodesic

Several illustrations have been given in Chapters I and II to show how the principle of least action can be applied to practical problems of dynamics, we give some further applications here.

Consider a particle P which moves on a smooth surface S but which is not acted on by any force other than the normal reaction of S. Since no work is done upon P, by the principle of conservation of energy its kinetic energy, and therefore its velocity, must be constant. From the statement of the principle of least action given by (3), § 5.6, it follows that $\int ds$ is stationary when integrated along the path of P. But the integral is stationary for a geodesic on S. Hence the path of P is a geodesic on S.

5.20. Dynamical paths related to geodesics on hypersurfaces

In a conservative dynamical system with parameters

$$q_1, q_2,..., q_n,$$

let the kinetic energy T be given by

$$T = \tfrac{1}{2}\sum_{i=1}^{n}\sum_{m=1}^{n} a_{im}\dot{q}_i\dot{q}_m \tag{1}$$

and let the potential energy be denoted by V. From (7), § 5.14, we have

$$T+V = h, \tag{2}$$

where h is constant. We also have the principle of least action which states that $2\int T\,dt$ is a minimum, or at least stationary, along a dynamical path.

From (1) we have

$$T^{\frac{1}{2}}\,dt = \Big(\tfrac{1}{2}\sum_{i=1}^{n}\sum_{m=1}^{n} a_{im}\,dq_i\,dq_m\Big)^{\frac{1}{2}}, \tag{3}$$

and from (2),

$$T^{\frac{1}{2}} = (h-V)^{\frac{1}{2}}. \tag{4}$$

Hence the principle of least action requires that along a dynamical path

$$2\int (h-V)^{\frac{1}{2}}\Big(\tfrac{1}{2}\sum_{i=1}^{n}\sum_{m=1}^{n} a_{im}\,dq_i\,dq_m\Big)^{\frac{1}{2}} \tag{5}$$

must be a minimum, or at least stationary. This is a geodesic on the hypersurface whose linear element ds is given by

$$ds^2 = 2(h-V)\Big(\sum_{i=1}^{n}\sum_{m=1}^{n} a_{im}\,dq_i\,dq_m\Big). \tag{6}$$

Thus to every trajectory corresponds a geodesic on the surface defined by (6). As an illustration consider the following example:

EXAMPLE 1. A particle describes a plane orbit under a central attractive force $\phi'(r)$, where r is its distance from the centre of force. Show that the orbit corresponds to a geodesic on the following surface of revolution: if (ρ,θ,z) are the cylindrical coordinates of a point on the surface and the z-axis is the axis of revolution, then $\rho^2 = r^2\{h-\phi(r)\}$ and the equation of the meridian curve is $z = f(\rho)$ where

$$f'(\rho) = \left\{\left(\frac{\rho\,dr}{r\,d\rho}\right)^2 - 1\right\}^{\frac{1}{2}}. \tag{7}$$

Here we have

$$-\frac{\partial V}{\partial r} = -\phi'(r) \quad \text{so that} \quad V = \phi(r).$$

Take (r,θ) to be the polar coordinates of the particle where θ is the azimuthal angle of the cylindrical coordinates. We have $T = \frac{1}{2}m(\dot r^2+r^2\dot\theta^2)$, so that by comparison with (1) above† we have $q_1 = r$, $q_2 = \theta$, $a_{11} = 1$, $a_{12} = 0$, $a_{22} = r^2$. From (6) we obtain the linear element of the surface whose geodesics

† A constant factor, such as m, can evidently be ignored.

correspond to the family of orbits of which one is described by the particle. It is

$$ds^2 = \{h-\phi(r)\}(dr^2+r^2\,d\theta^2). \tag{8}$$

Now, Fig. V. 3, the linear element on a surface of revolution

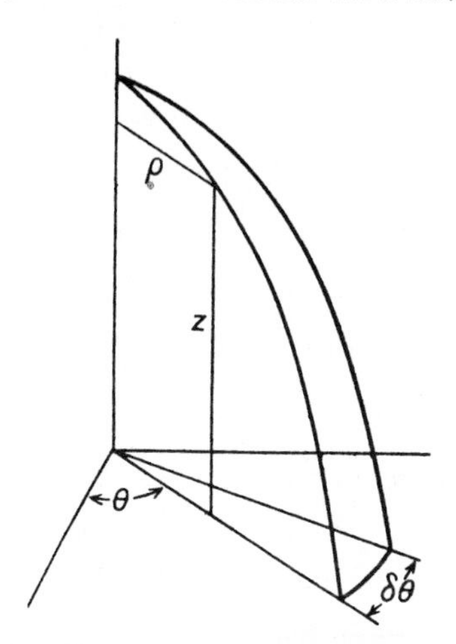

FIG. V. 3.

with the z-axis as the axis of revolution and with (ρ, θ, z) as cylindrical coordinates is given by

$$ds^2 = d\rho^2+dz^2+\rho^2\,d\theta^2. \tag{9}$$

Comparison of (8) and (9) shows that

$$\rho^2 = r^2\{h-\phi(r)\} \tag{10}$$

and that

$$d\rho^2+dz^2 = \{h-\phi(r)\}\,dr^2 \tag{11}$$

$$= \frac{\rho^2\,dr^2}{r^2} \tag{12}$$

on using (10). From (12) we have

$$\frac{dz}{d\rho} = \left\{\left(\frac{\rho\,dr}{r\,d\rho}\right)^2-1\right\}^{\frac{1}{2}}, \tag{13}$$

so that, if $z = f(\rho)$, then

$$f'(\rho) = \left\{\left(\frac{\rho\,dr}{r\,d\rho}\right)^2-1\right\}^{\frac{1}{2}} \tag{14}$$

as required.

EXAMPLE 2. A particle acted on by a constant gravitational force describes a parabola. Show that such parabolic orbits correspond to geodesics on the following surface of revolution. Let x and y, respectively, be the horizontal and vertical coordinates of the particle at time t and let (ρ, θ, z) be the cylindrical coordinates of the corresponding point on the surface of revolution, where the z-axis is the axis of revolution. Then

$$\rho = a(h-2gy)^{\frac{1}{2}}, \qquad \theta = x/a,$$

and along the meridian

$$\frac{dz}{d\rho} = \left(\frac{\rho^4}{a^6 g^2} - 1\right)^{\frac{1}{2}},$$

where g is the constant of gravity and a and h are constants.

5.21. Hamilton's equations

Other systems of dynamical equations exist which are as comprehensive in their scope as the Lagrange equations, for example the canonical equations of Hamilton. These equations, which can be derived from Lagrange's, have important applications to problems in the Calculus of Variations as well as to dynamics. We shall first derive them from Lagrange's equations and then illustrate their use in the calculus.

With the terminology already used in this chapter let

$$L = T - V = L(q_1, q_2,..., q_n; \dot{q}_1, \dot{q}_2,..., \dot{q}_n; t) \tag{1}$$

be the kinetic potential and let

$$p_i = \frac{\partial L}{\partial \dot{q}_i} \quad (i = 1, 2,..., n) \tag{2}$$

$$= \frac{\partial T}{\partial \dot{q}_i}, \tag{3}$$

since the potential is a function of the parameters q_i only.

Let H, the Hamiltonian function, or simply the Hamiltonian, be defined by the equation

$$H = -L + \sum_{i=1}^{n} p_i \dot{q}_i. \tag{4}$$

From (2) we may solve for $\dot{q}_i$ $(i = 1, 2,..., n)$, and then eliminate $\dot{q}_i$ from (4), leaving H as a function of q_i, p_i, and t. Or, what is essentially the same, we may regard H as a function of q_i, $\dot{q}_i$, and t, where $\dot{q}_i$ is a function of p_m, q_m $(m = 1, 2,..., n)$, and t. Taking the latter view we have

$$\frac{\partial H}{\partial p_i} = -\sum_{m=1}^{n} \frac{\partial L}{\partial \dot{q}_m}\frac{\partial \dot{q}_m}{\partial p_i} + \dot{q}_i + \sum_{m=1}^{n} p_m \frac{\partial \dot{q}_m}{\partial p_i} \tag{5}$$

$$= \dot{q}_i, \tag{6}$$

from (2). Also

$$\frac{\partial H}{\partial q_i} = -\frac{\partial L}{\partial q_i} - \sum_{m=1}^{n} \frac{\partial L}{\partial \dot{q}_m}\frac{\partial \dot{q}_m}{\partial q_i} + \sum_{m=1}^{n} p_m \frac{\partial \dot{q}_m}{\partial q_i} \tag{7}$$

$$= -\frac{\partial L}{\partial q_i} \tag{8}$$

from (2), $$= -\frac{d}{dt}\left(\frac{\partial L}{\partial \dot{q}_i}\right) \tag{9}$$

from Lagrange's equations, (2), § 5.7,

$$= -\dot{p}_i, \tag{10}$$

from (2) again. Equations (6) and (10) $(i = 1, 2,..., n)$ comprise Hamilton's equations and from them p_i, q_i $(i = 1, 2,..., n)$ can be determined as functions of t. They are frequently referred to as Hamilton's canonical equations.

In the case of conservative systems, H, the Hamiltonian function has a simple interpretation. For T is now a homogeneous quadratic function of $\dot{q}_i$, § 5.14, and so by Euler's theorem on homogeneous functions we have, from (3),

$$\sum_{i=1}^{n} p_i \dot{q}_i = \sum_{i=1}^{n} \frac{\partial T}{\partial \dot{q}_i} \dot{q}_i = 2T. \tag{11}$$

Hence

$$H = -L + \sum_{i=1}^{n} p_i \dot{q}_i = -T + V + 2T = T + V, \tag{12}$$

and H is thus equal to the total energy.

The theory developed so far deals with Hamilton's equations largely from the dynamical point of view. But it is evident that even if $\int L\,dt$ does not arise from a dynamical problem,

Hamilton's equations (6) and (10) are still equivalent to the characteristic equations

$$\frac{d}{dt}\left(\frac{\partial L}{\partial \dot{q}_i}\right)-\frac{\partial L}{\partial q_i}=0. \tag{13}$$

Hamilton's equations can be used in the calculus of variations independently of dynamics, although in such cases we cannot associate H with the concept of energy. This will be illustrated in § 5.23 by the discussion of a special case of Zermelo's navigational problem. In the next paragraph we shall deal with some interesting properties of the Hamiltonian function and equations for the non-dynamical case where L is a homogeneous function of degree one in the variables $\dot{q}_i$ $(i = 1,..., n)$.

5.22. The non-dynamical case when L is homogeneous and of degree one in $\dot{q}_i$ $(i = 1, 2,..., n)$

The case where L is of degree one in $\dot{q}_i$ cannot arise in dynamical theory since in dynamics L must be of degree two in $\dot{q}_i$. It is, however, frequently encountered in geometrical and similar problems and has many features of interest.

In this section we assume that L is homogeneous of degree one in $\dot{q}_i$ $(i = 1, 2,..., n)$, and (ii) that L does not contain t explicitly. We shall express the dependence of L on the q's and $\dot{q}$'s by writing

$$L = L(q_i, \dot{q}_i) = L(q_1, q_2,..., q_n; \dot{q}_1, \dot{q}_2,..., \dot{q}_n). \tag{1}$$

By Euler's theorem on homogeneous functions we have

$$\dot{q}_1\frac{\partial L}{\partial \dot{q}_1}+\dot{q}_2\frac{\partial L}{\partial \dot{q}_2}+...+\dot{q}_n\frac{\partial L}{\partial \dot{q}_n} = L. \tag{2}$$

In finding the stationary values of $\int L\,dt$ by means of Hamilton's equations we first find the functions p_i defined by

$$p_i = \frac{\partial L}{\partial \dot{q}_i} \quad (i = 1, 2,..., n) \tag{3}$$

and then eliminate $\dot{q}_i$ $(i = 1, 2,..., n)$ from the function H defined by

$$H = p_1\dot{q}_1+p_2\dot{q}_2+...+p_n\dot{q}_n-L, \tag{4}$$

as in § 5.21. From (2) and (3) it follows that in the case considered here $H = 0$. Or, writing H in a form in which it is expressed as a function of q_i and p_i $(i = 1, 2,..., n)$ we have

$$H(q_1, q_2,..., q_n; p_1, p_2,..., p_n) = 0. \tag{5}$$

The converse of this result is also true, as we now proceed to show. The converse can be stated as follows:

Let $L = L(q_i, \dot{q}_i)$ be a homogeneous function of degree one in $\dot{q}_i$ $(i = 1, 2,..., n)$, and let p_i be defined by (3). If between the variables q_i, p_i there exists a relationship given by

$$K = K(q_1, q_2,..., q_n; p_1, p_2,..., p_n) = 0, \tag{6}$$

then K is the Hamiltonian function for L.

For let F be the function whose Hamiltonian is K. Then from the definition of the Hamiltonian given by (4), § 5.21, and the condition that p_i is defined by equation (3), we have

$$K = \sum_{i=1}^{n} \frac{\partial L}{\partial \dot{q}_i} \dot{q}_i - F. \tag{7}$$

On using (2) and (6) it follows immediately that $F = L$ and therefore that L has K as its Hamiltonian function.

The canonical equations (6) and (10), § 5.21, which render $\int L\,dt$ stationary, are then

$$\frac{\partial K}{\partial p_i} = \dot{q}_i, \tag{8}$$

and

$$\frac{\partial K}{\partial q_i} = -\dot{p}_i \quad (i = 1, 2,..., n). \tag{9}$$

This result is sometimes of great use in simplifying our calculations, for frequently it is easier to find a relation such as (6) than to perform the eliminations necessary to find H by the methods of § 5.21. This point is illustrated by (15), § 5.23, below.

Another interesting property of the Hamiltonian in the case when L is homogeneous and of degree one in the variables $\dot{q}_i$ is as follows. Having found the Hamiltonian H according to the rules of § 5.21, the solutions of the canonical equations

$$\frac{\partial H}{\partial p_i} = \dot{q}_i; \qquad \frac{\partial H}{\partial q_i} = -\dot{p}_i \quad (i = 1, 2,..., n), \tag{10}$$

are the extremals of $\int L\,dt$.

Now write $t = f(u)$ and denote differentiation with respect to u by a prime, i.e. $q_i' = dq_i/du$, etc. If L is homogeneous of degree one in $\dot{q}_1$ and if in addition t is not explicit in L, we have

$$L(q_1, q_2,..., q_n; \dot{q}_1, \dot{q}_2,..., \dot{q}_n) = \frac{1}{f'(u)} L(q_1, q_2,..., q_n; q_1', q_2',..., q_n'), \quad (11)$$

so that

$$\int L(q_1, q_2,..., q_n; \dot{q}_1, \dot{q}_2,..., \dot{q}_n)\, dt = \int L(q_1, q_2,..., q_n; q_1', q_2',..., q_n')\, du. \quad (12)$$

Also the Hamiltonian equations (10) become

$$f'(u)\frac{\partial H}{\partial p_i} = q_i'; \qquad f'(u)\frac{\partial H}{\partial q_i} = -p_i' \quad (i = 1, 2,..., n). \quad (13)$$

Thus the right-hand integral of (12) is stationary for solutions of (13) and on replacing u by t (the actual symbol used is immaterial), it follows that $\int L\, dt$ is stationary for solutions of the equations

$$\mu\frac{\partial H}{\partial p_i} = \dot{q}_i; \qquad \mu\frac{\partial H}{\partial q_i} = -\dot{p}_i \quad (i = 1, 2,..., n), \quad (14)$$

where μ is any convenient function of t. By choosing μ suitably our calculations can frequently be simplified.

These results will be illustrated in the next section, where we deal with a special case of Zermelo's navigational problem.† We shall confine ourselves to the first variation only.

5.23. Path of minimum time in a stream with given flow

A stream of water flows parallel to a fixed given horizontal line σ in such a manner that its velocity at a point distant q_2 from σ is proportional to q_2. If a ship moves with constant speed u relative to the water, find the path of minimum time between two points in the stream.

Let (q_1, q_2) be the coordinates of the ship at time t, where q_1 is measured parallel to σ, and let ϕ be the angle between σ and the direction in which the ship is being steered. Then the

† E. Zermelo, *Zeitschrift für angewandte Mathematik und Mechanik*, vol. xi (1931), pp. 114–24. A full account is given by C. Carathéodory, *Variationsrechnung*, pp. 276 and 458.

components of the ship's velocity relative to σ are respectively $kq_2+u\cos\phi$ and $u\sin\phi$ parallel to the q_1 and q_2 axes. By suitable choice of units it is possible to make $k = 1$ and $u = 1$, so simplifying these components to $q_2+\cos\phi$ and $\sin\phi$ respectively.

If v is the velocity of the ship relative to σ and s the length of arc of its trajectory, then $v = ds/dt$, so that the integral to be minimized is

$$t = \int \frac{ds}{v}. \tag{1}$$

In order to put this integral in a form to which the theory of § 5.22 can be applied a parameter λ is introduced. The object of this is to replace the variable t and so avoid possible confusion with the time. Denote differentiations with respect to λ by dashes, e.g. $s' = ds/d\lambda$, etc., so that

$$t = \int \frac{ds}{v} = \int \frac{s'}{v}\,d\lambda. \tag{2}$$

The integrand s'/v must now be expressed in the form

$$L(q_1, q_2;\, q_1', q_2';\, \lambda) \quad (= L \text{ for brevity}).$$

We then have

$$\frac{s'}{v} = L, \tag{3}$$

where v is the velocity of the ship relative to the fixed line and L remains to be determined.

If the direction of motion of the ship makes an angle ψ with the given line σ, then, by the usual formulae of the differential calculus,

$$\tan\psi = \frac{\sin\phi}{q_2+\cos\phi}; \qquad \sin\psi = \frac{dq_2}{ds}; \qquad \cos\psi = \frac{dq_1}{ds}. \tag{4}$$

Hence

$$v\cos\psi = v\frac{dq_1}{ds} = v\frac{q_1'}{s'} = \frac{q_1'}{L}, \tag{5}$$

from (3). Similarly

$$v\sin\psi = v\frac{dq_2}{ds} = \frac{q_2'}{L}. \tag{6}$$

Now the velocity components can be written either as

$$(v\cos\psi,\, v\sin\psi) \quad \text{or as} \quad (q_2+\cos\phi,\, \sin\phi).$$

Hence we have $$q_2+\cos\phi = \frac{q_1'}{L}, \tag{7}$$

and $$\sin\phi = \frac{q_2'}{L}. \tag{8}$$

On eliminating ϕ from equations (7) and (8) we have the quadratic equation
$$(Lq_2-q_1')^2+q_2'^2 = L^2, \tag{9}$$
from which L can be found. We can then minimize (2) by using the characteristic equations. But, as we wish to illustrate the use of Hamilton's equations and as, evidently from (9), L is homogeneous and of degree one in q_1' and q_2', we shall proceed according to the methods of §§ 5.21 and 5.22. It is first necessary to calculate the two quantities p_1 and p_2, where
$$p_1 = \frac{\partial L}{\partial q_1'} \quad \text{and} \quad p_2 = \frac{\partial L}{\partial q_2'}, \tag{10}$$
and this is most conveniently achieved by differentiating (9) with respect to q_1' and q_2'. We then obtain the equations
$$(p_1q_2-1)(Lq_2-q_1') = Lp_1, \tag{11}$$
$$p_2q_2(Lq_2-q_1')+q_2' = Lp_2. \tag{12}$$
Using (7) and (8) these equations can be rewritten in the form
$$(p_1q_2-1)(-\cos\phi) = p_1, \tag{13}$$
$$p_2q_2(-\cos\phi)+\sin\phi = p_2, \tag{14}$$
and on eliminating ϕ we then have
$$p_1^2+p_2^2-(p_1q_2-1)^2 = 0. \tag{15}$$
From (9) L is homogeneous and of degree one in q_1' and q_2'; hence from the results of § 5.22 the Hamiltonian H is given by
$$H = p_1^2+p_2^2-(p_1q_2-1)^2. \tag{16}$$
From (14), § 5.22, the canonical equations take the form
$$\mu\{p_1-q_2(p_1q_2-1)\} = q_1', \tag{17}$$
$$\mu p_2 = q_2', \tag{18}$$
$$0 = -p_1', \tag{19}$$
$$-\mu\{p_1(p_1q_2-1)\} = -p_2', \tag{20}$$
where μ is a function of λ which can be chosen arbitrarily.

From (19) we have $p_1 = 1/c$, where c is a constant independent of λ. Choose μ to be constant, differentiate (18), and substitute for p_2' in (20). We obtain

$$c^2 q_2'' = \mu^2(q_2 - c). \tag{21}$$

Now choose μ so that

$$\mu^2 = c^2. \tag{22}$$

The solution of (21) is then

$$q_2 = c + Ae^{\lambda} + Be^{-\lambda}, \tag{23}$$

where A and B are arbitrary constants.

From (18) we have

$$\mu p_2 = q_2' = Ae^{\lambda} - Be^{-\lambda}, \tag{24}$$

and substituting for p_1, p_2, and q_2 in (15) and using (22) we have $AB = \frac{1}{4}$. But the parameter λ is a function of the time, and on measuring the time so that $\lambda = 0$ at the instant when $q_2' = 0$ it follows from (24) that $A = B$. Finally then, $A^2 = \frac{1}{4}$ and $A = \pm\frac{1}{2}$. For the rest of the paragraph whenever an ambiguous sign occurs the upper sign is to be taken if $A = \frac{1}{2}$ and the lower sign if $A = -\frac{1}{2}$. We then have

$$q_2 = c \pm \cosh\lambda; \qquad \mu p_2 = \pm\sinh\lambda. \tag{25}$$

Since L is homogeneous and of degree one in q_1' and q_2' we have

$$L = p_1 q_1' + p_2 q_2' = \mp(\mu\cosh\lambda)/c \tag{26}$$

on using the values of p_1, p_2, q_1', q_2', and μ given by (17), (18), (19), and (20). The sign of μ can be chosen arbitrarily and the choice is made so that L is always positive. It is for this purpose that μ has been introduced. We then have

$$\mp\mu/c = 1 \tag{27}$$

and $L = \cosh\lambda$. From (2) and (3) we have

$$t = \int_0^{\lambda} L\, d\lambda = \sinh\lambda. \tag{28}$$

All the parameters can now be expressed in terms of the time, [q_1 after integrating (17)]. The final results are

$$q_1 - d = ct \pm \tfrac{1}{2}[t(1+t^2)^{\frac{1}{2}} - \log_e\{t + (1+t^2)^{\frac{1}{2}}\}], \tag{29}$$

$$q_2 = c \pm (1+t^2)^{\frac{1}{2}}, \tag{30}$$

$$p_1 = (1/c), \tag{31}$$

$$p_2 = -(t/c), \tag{32}$$

where c and d are arbitrary constants. In the case of ambiguity of sign, all upper or all lower signs are taken together. The coordinates of the ship at time $t = 0$ are $(d, c \pm 1)$ where the ambiguous sign obeys the rule above.

The formulae obtained for $\cos\phi$ and $\sin\phi$ on substituting for q_1', q_2', q_2, and L in (7) and (8) are

$$\cos\phi = \mp\frac{1}{(1+t^2)^{\frac{1}{2}}}; \qquad \sin\phi = \pm\frac{t}{(1+t^2)^{\frac{1}{2}}}, \tag{33}$$

where again the ambiguous signs obey the same rule as those of equations (29) et seq. Equations (33) show that the ship can be steered along the path of minimum time by blind reckoning since the angle ϕ can be calculated at any instant independently of external observations.

Interesting diagrams of the paths and a detailed discussion of the question whether maxima or minima occur can be found in the work of Carathéodory† already referred to.

EXAMPLE 1. Solve by the above method the problem of the Brachistochrone, § 1.11. (The integral to be minimized is

$$\int \frac{(\dot{q}_1^2+\dot{q}_2^2)^{\frac{1}{2}}}{q_2^{\frac{1}{2}}}\,dt,$$

where q_1 and q_2 are respectively the horizontal and vertical coordinates of the particle, and $H \equiv p_1^2+p_2^2-(1/q_2) = 0$. Use (6) and (10), § 5.21.)

EXAMPLE 2. Solve by the above method the problem of the minimum surface of revolution, § 1.12. (The integral to be minimized is

$$\int q_2(\dot{q}_1^2+\dot{q}_2^2)^{\frac{1}{2}}\,dt,$$

where q_1 and q_2 are suitably chosen Cartesian coordinates, and

$$H \equiv p_1^2+p_2^2-q_2^2 = 0.)$$

† p. 242 et seq.

CHAPTER VI

HAMILTON'S PRINCIPLE IN THE SPECIAL THEORY OF RELATIVITY

6.1. Introduction

It is clear from the results obtained in the previous chapters, and particularly from Chapter V, that configurations for which energy functions have stationary or minimum values are of great importance in the study of physical phenomena. The basic principles of such phenomena can be expressed more concisely and the mathematical analysis effected more readily by using the calculus of variations than by any other means.

Hamilton's stationary principle plays an important part in the theory of relativity, especially in the generalized theory,† but an account of this theory is beyond the scope of any work which does not establish the Tensor Calculus first. In this chapter we shall only deal with the special theory of relativity and show how the stationary character of $\int L\,dt$, which plays such an important part in Newtonian dynamics, can lead us naturally from Newtonian to Einsteinian mechanics.

6.2. The physical bases of the special theory of relativity

The wave theory of light, advocated by Fresnel and Young early in the nineteenth century, conceived the ether as the medium in which the undulations of light take place. Galilean dynamics precludes the possibility of absolute velocity, since laws of nature referred to a frame G_1 still remain true when referred to a frame G_2 whose velocity relative to G_1 is uniform and translatory. But we are not prevented from choosing any convenient frame and setting up our dynamics relative to it. Since the ether, sometimes referred to as the stagnant ether, naturally suggests itself as a most convenient frame of reference numerous experiments were performed in order to determine

† A. S. Eddington, *The Mathematical Theory of Relativity*, § 60, pp. 137 et seq., H. Weyl, *Space, Time, and Matter*, §§ 28 and 36.

the velocity of the earth relative to it. Some of these experiments were based on optical and some on electrical methods, but the results were always the same; the velocity of the earth relative to the ether is zero.

After much controversy the first convincing explanation was given by Einstein† in 1905, an explanation so comprehensive that a number of hitherto inexplicable facts were accounted for in addition to the null results of these experiments. Before discussing the theory we give a slight account of two of the most famous experiments, that of Michelson and Morley,‡ performed in 1887, and that of Trouton and Noble,§ performed in 1903. Both experiments were repeated with greater refinements only to confirm the result that the velocity of the earth relative to the ether is always zero.

6.3. The Michelson and Morley experiment

A ray of light, emitted from a source A (Fig. VI. 1), strikes a partly silvered plane glass sheet G at 45° to the normal and is divided into two rays r_1 and r_2, r_1 being reflected and r_2 transmitted. The rays r_1 and r_2 are reflected back to G by two mirrors M_1 and M_2, r_1 being then transmitted through G and r_2 reflected from G. The two rays r_1 and r_2 are now recombined along the line GB, after each part has experienced one reflection and one transmission at G. If there is a difference of time between the two paths from A to B there will be a consequent difference of phase between the two parts of the ray when recombined and this can be detected by the usual technique of interference patterns.

The two paths of light as seen by an observer moving with the apparatus are shown in Fig. VI. 1 and the paths as seen by an observer at rest relative to the ether are shown in Fig. VI. 2. In Fig. VI. 2 the various parts of the apparatus are not shown at the same instant of time, for example during the passage of time from reflection at G to M_1 and back the glass sheet G has moved from G_1 to G_2. Fig. VI. 2 raises the question of how light is

† A. Einstein, 'Zur Elektrodynamik bewegter Körper', *Annalen der Physik*, **17** (1905).

‡ *Phil. Mag.* Dec. 1887.

§ *Proc. Roy. Soc.* **72** (1903), p. 132; *Phil. Trans.* **202** (1903), p. 165.

reflected from moving mirrors. This is easily answered by means of Huygens's principle and is dealt with in most books on physics.†

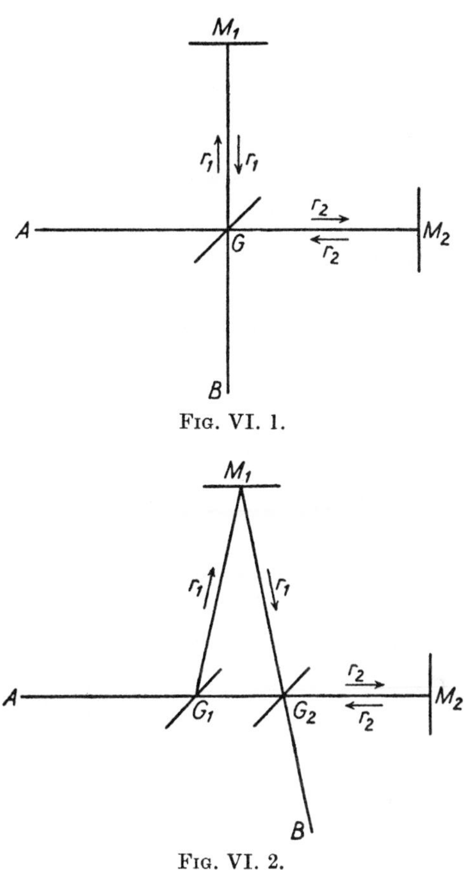

FIG. VI. 1.

FIG. VI. 2.

Suppose that the apparatus is moving in the direction of GM_2 with velocity v relative to the ether and that c denotes the speed of light.‡ Then during the journey from G to M_2 the velocity of light relative to the apparatus is $c-v$ and from M_2 back to G it is $c+v$. The time from G to M_2 and back is therefore

$$\frac{GM_2}{c-v}+\frac{GM_2}{c+v}=\frac{2cGM_2}{c^2-v^2}. \tag{1}$$

† See E. Cunningham, *The Principle of Relativity*, pp. 19 and 20 for a discussion of this point.

‡ $c = 3.10^{10}$ cm. per second.

Along the path GM_1 and back the velocity of light relative to the apparatus is $\sqrt{(c^2-v^2)}$ and the time taken is therefore

$$\frac{2GM_1}{\sqrt{(c^2-v^2)}}. \tag{2}$$

Hence the excess of the time along the path to M_2 and back over that to M_1 and back is

$$\frac{2cGM_2}{c^2-v^2}-\frac{2GM_1}{\sqrt{(c^2-v^2)}}. \tag{3}$$

If the apparatus is turned through a right angle so that GM_1 now lies along the direction of v, its velocity relative to the ether, then the same argument shows that the excess of the time along the path GM_2G over the path GM_1G is

$$\frac{2GM_2}{\sqrt{(c^2-v^2)}}-\frac{2cGM_1}{c^2-v^2}. \tag{4}$$

The difference between expressions (3) and (4) is

$$2(GM_1+GM_2)\left\{\frac{c}{c^2-v^2}-\frac{1}{\sqrt{(c^2-v^2)}}\right\} = (GM_1+GM_2)\frac{v^2}{c^3}, \tag{5}$$

when terms of order smaller than (v^2/c^3) are neglected.

Although the right-hand side of (5) is extremely small, yet, by means of the technique of interference, it would have been detected even if v were only one-tenth of the earth's velocity relative to the sun. It requires little consideration to realize that during its orbit round the sun the earth's velocity relative to the ether must at some point be at least as great as its velocity relative to the sun. It therefore came as a great surprise when the first experiments revealed that $v = 0$. Repetition of these experiments, with every possible refinement, confirmed this result conclusively.

6.4. The Trouton and Noble experiment

This experiment had the same object as the Michelson and Morley experiment, namely the determination of the velocity of the earth relative to the ether, but the method adopted was entirely different.

A moving electric charge generates a current and with each current, as discovered by Oersted, is associated a magnetic field. Thus if a charged parallel plate condenser is at rest in the ether it possesses electrostatic energy only. But if it is in motion the currents generated by the equal and opposite charges of the condenser plates set up a magnetic field whose energy must be added to that of the electrostatic field. Hence the total electromagnetic energy of the condenser is a function of its velocity relative to the ether.

Let T denote the total electromagnetic energy, which is not difficult to calculate, and let ψ denote the angle between the direction of velocity of the condenser relative to the ether and the normal to the plates. Then by standard mechanical theory $-(\partial T/\partial\psi)$ is equal to the moment of the couple tending to increase the value of ψ. On evaluation† it is found that the moment of this couple is proportional to v^2/c^2, where v is the velocity of the condenser relative to the ether and c, as before, denotes the speed of light. Hence if the condenser is suspended by a torsion thread and charged and discharged alternately it should rotate through an angle proportional to v^2.

Although the experiment was performed with great accuracy, the value of v deduced was again zero in every case.

6.5. The principles of special relativity

The null results of these experiments indicate clearly that the speed of light must be the same for all observers independently of their individual velocities. This statement is one of the fundamental postulates of the special theory of relativity.

The following considerations show how widely this postulate and its consequences differ from Galilean or Newtonian ideas. Let the position of a point in space be determined by Cartesian coordinates (x, y, z), the origin of coordinates being O; we shall refer to this as the O system. Suppose (i) that an observer M moves along the x-axis with constant velocity v and (ii) that at the instant when M passes O a source of light at O instantaneously emits rays in all directions with velocity c relative to

† For the calculations see the references in the footnote to § 6.3.

the axes. Then after time t the equation of the spherical wave front formed by the light rays will be

$$x^2+y^2+z^2-c^2t^2 = 0, \tag{1}$$

and the coordinates of M will be $(vt, 0, 0)$ (Fig. VI. 3).

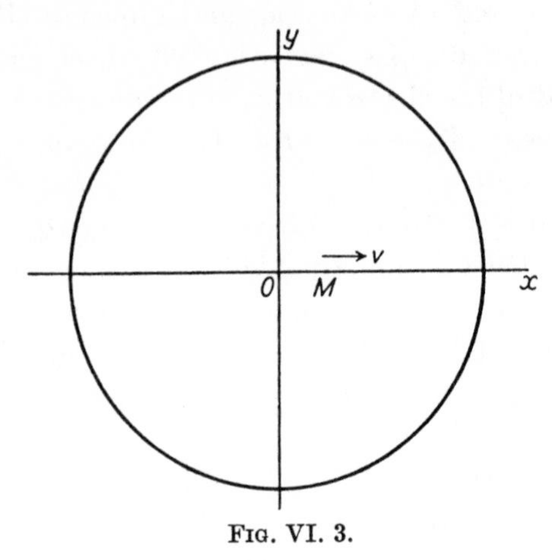

FIG. VI. 3.

Consider now the system of axes which are parallel to those through O but which have M as origin; we shall refer to this as the M system and denote the coordinates of a point in it by (x', y', z'). With respect to the M system the wave front is still a sphere of radius ct, but its centre has coordinates $(-vt, 0, 0)$ and its equation is

$$(x'+vt)^2+y'^2+z'^2-c^2t^2 = 0. \tag{2}$$

But clearly, if we adopt Einstein's postulate stated at the beginning of this section, the distance of the wave front from M must be ct in all directions, for the speed of light relative to M is always c independently of M's motion. The equation of the wave front in the M system must then be

$$x'^2+y'^2+z'^2-c^2t^2 = 0 \tag{3}$$

instead of (2).

Reverting to Galilean conceptions of relativity the linear relations which enable us to change from the O to the M system are those which transform (1) to (2) and vice versa. They are

$$\left.\begin{aligned} x &= x'+vt, \quad y = y', \quad z = z', \\ x' &= x-vt, \quad y' = y, \quad z' = z. \end{aligned}\right\} \tag{4}$$

Galilean theory asserts that all physical phenomena must appear the same whether viewed from a frame of reference G_1 or a frame G_2, provided that either frame moves with uniform velocity relative to the other. If these phenomena are expressible in mathematical form, then such forms must be covariant† for transformation (4).

Starting from the initial postulate of this section, the Einstein theory requires our laws of relativity to be based upon the linear transformations which change (1) to (3) instead of (1) to (2) (the transformations which change (1) to (3) will be obtained in § 6.7 below). The Einstein theory thus requires that all laws of physics expressible in mathematical form must be covariant for (3), § 6.7, instead of for (4) above.

In order to make further progress in his theory Einstein introduced another very remarkable postulate. Before stating it we note that in Newtonian theory the space variables are dealt with in quite a different way from the time variable t, which appears to be specially privileged.‡ Although two observers, one in the O and one in the M system, assign different values to the coordinates and relative velocities of a particle P, yet they assign the same value to t. In this concept time is independent of the velocity of any observer, as illustrated by the use of t in (4). Although this agrees with our intuitive conceptions it is rejected by Einstein, who assumes that the measurement of time depends as much upon the velocity of the observer as does the measurement of velocity. Thus in passing from the O to the

† *Definition of covariant.* If a transformation from the variables $x_1, x_2,..., x_n$ to $y_1, y_2,..., y_n$ is given by $x_i = f_i\,(y_1, y_2,..., y_n)$ $(i = 1, 2,..., n)$, then the function $F(x_1, x_2,..., x_n)$ is a covariant of the transformation if

$$F(x_1, x_2,..., x_n) = F(y_1, y_2,..., y_n).$$

As an example, for the group of rotations about the origin O given by

$$x_1 = y_1 \cos\alpha - y_2 \sin\alpha, \qquad x_2 = y_1 \sin\alpha + y_2 \cos\alpha$$

the expression $x_1^2+x_2^2$ is covariant since $x_1^2+x_2^2 = y_1^2+y_2^2$. Interpreted geometrically, if the point P rotates about O, then the distance OP remains unaltered.

The essential property of a covariant for our purpose is that its form remains unaltered by a transformation. This is wider than the normal usage of the word in generalized relativity theory.

‡ Or perhaps it is the space variables which are privileged.

M system not only must we make the change from (x, y, z) to (x', y', z') but we must also allow for a possible change of time measurement from t to t'. This postulate proves most fertile in its applications to physical problems.

Since the space variables (x, y, z) and the time variable t are now on an equal basis, it is natural to group them together and by analogy with three-dimensional geometry to regard the set of numbers (x, y, z, t) as the coordinates of a point in a four-dimensional space. This is generally referred to as space-time (or the space-time continuum). The second postulate states that if (x, y, z, t) are the space-time coordinates of a point P in the O system, then in the M system its space-time coordinates will be (x', y', z', t'), where t' differs from t because of M's velocity relative to O. Thus the equation of the wave front in the M system must be

$$x'^2+y'^2+z'^2-c^2t'^2 = 0 \tag{5}$$

instead of (3) (t is replaced by t'). To sum up, as a consequence (i) of the null results of the experiments described in §§ 6.3 and 6.4 and (ii) of the displacement of time from its privileged position accorded in Galilean concepts the relativity laws must be linear transformations which change (1) to (5) or, in other words, which leave $x^2+y^2+z^2-c^2t^2$ invariant.

6.6. Galilean and Newtonian conceptions of time

Before considering the consequences of these postulates it is worth while considering the concept of time according to the ideas of Galileo and Newton. For although simple and intuitive it is by no means free of paradox, as the following example will show.

Suppose that two observers A and B correlate their clocks when together and that c cm. per second is the speed of light. Let A move to a point distant c cm. from B and then come to rest relative to B. According to Newtonian conceptions these changes do not affect A's time system, i.e. his clock, in any manner. Now if A and B are at rest relative to the ether and if light waves are the only medium of communication, then B's clock will appear to be one second slow to A and vice versa.

But if A and B are not at rest relative to the ether then the apparent time differences of their clocks will depend upon their velocity. If they are moving in the direction B to A, then B's clock will appear to be more than one second slow to A and A's clock will appear to be less than one second slow to B, and if they are moving in the opposite direction then the contrary will appear to be true. Suppose now that A wishes to determine how events which appear simultaneous to him will appear to B. Then he must first find his velocity relative to the ether and this, according to the Michelson and Morley experiment, is impossible. Thus the so-called intuitive conceptions of time give us no means for correlating the time systems of two observers and in this respect the Einstein theory is undoubtedly superior.†

6.7. The transformations of the special theory of relativity

Reverting to the theory of § 6.5, we have already assumed that the transformations of relativity must be linear, an assumption justified by Einstein because of the homogeneity of space and time. In addition a further assumption is made that a point at rest in one system or frame of reference must correspond to a point moving with uniform velocity in the other. Evidently $y = y'$, $z = z'$, so that the transformations from the O to the M systems of § 6.5 must be of the form

$$x = \beta(x'+vt'); \quad y = y'; \quad z = z'; \quad t = px'+qy'+rz'+st'. \tag{1}$$

The constants β, p, q, r, s must be determined so that (1), § 6.5 is transformed to (5), § 6.5. We have

$$\beta^2(x'+vt')^2+y'^2+z'^2-c^2(px'+qy'+rz'+st')^2 = x'^2+y'^2+z'^2-c^2t'^2, \tag{2}$$

and so $q = r = 0$,

$$x = \beta(x'+vt'): \quad y = y'; \quad z = z'; \quad t = \beta\left(t'+\frac{vx'}{c^2}\right), \tag{3}$$

where
$$\beta = \left(1-\frac{v^2}{c^2}\right)^{-\frac{1}{2}}. \tag{4}$$

† In his book *Space, Time, and Matter*, p. 147, H. Weyl says: 'The physical purport of this is that we are to discard our belief in the objective meaning of simultaneity; it was the great achievement of Einstein in the field of the theory of knowledge that he banished this dogma from our minds, and this is what leads us to rank his name with that of Copernicus'.

The reverse transformation from the M to the O system is

$$x' = \beta(x-vt); \qquad y' = y; \qquad z' = z; \qquad t' = \beta\left(t-\frac{vx}{c^2}\right), \quad (5)$$

where β has the value given by (4).

The interesting feature of these transformations is that not only do they account for the null results of the Michelson and Morley and similar experiments but that they had already been in use some years prior to the publication of Einstein's theory in order to account for an entirely different phenomenon. The study of electricity and magnetism in the early part of the nineteenth century led finally to a system of equations, known as Maxwell's equations. These cover the whole field of electromagnetic phenomena and are so completely in accord with experiment that they have become the classical equations of electromagnetic theory. Nevertheless they suffered from one serious drawback, they were not covariant for the Galilean relativity transformations given by (4), § 6.5. This difficulty provoked much thought and study until it was ultimately discovered by Lorentz and Larmor that Maxwell's equations are covariant for the transformations (3) and (5) above.† This remarkable success in explaining two groups of apparently unconnected phenomena together with many other successes, some of which we shall deal with later, have firmly established Einstein's special theory of relativity in the field of modern mathematical physics.

6.8. Relativity transformations for small time intervals

In § 6.7 it has been assumed that v is constant. If we wish to extend the theory to the case when v is no longer constant we must consider time intervals so small that for their duration the variations of v are negligible. In the formulae of § 6.7 we replace (x, y, z, t) in the O system of reference by (dx, dy, dz, dt) and (x', y', z', t') in the M system by (dx', dy', dz', dt'), and the

† For the proof of this statement see E. Cunningham, loc. cit., chap. iv, or R. C. Tolman, *Relativity Thermodynamics and Cosmology*, § 52. The electrical and magnetic intensities also undergo transformations similar to (2) above.

theory then requires that the relativity transformations from the O to the M system must change

$$dx^2+dy^2+dz^2-c^2\,dt^2 \tag{1}$$

to

$$dx'^2+dy'^2+dz'^2-c^2\,dt'^2. \tag{2}$$

The transformations (3) and (5), § 6.7, from the O to the M system of reference and back, then become

$$\left.\begin{aligned} dx &= \beta(dx'+v\,dt'), \quad dy = dy', \quad dz = dz', \quad dt = \beta\left(dt'+\frac{v\,dx'}{c^2}\right), \\ dx' &= \beta(dx-v\,dt), \quad dy' = dy, \quad dz' = dz, \quad dt' = \beta\left(dt-\frac{v\,dx}{c^2}\right), \end{aligned}\right\} \tag{3}$$

where

$$\beta = \left(1-\frac{v^2}{c^2}\right)^{-\frac{1}{2}}.$$

6.9. The space-time continuum

We have already mentioned that it is convenient to group the space and time variables together and to regard (x,y,z,t) as the coordinates of a point in four-dimensional space, referred to as the space-time continuum. It is more convenient to take (x,y,z,ict) $\{i = \sqrt{(-1)}\}$ as the coordinates of the point so that

$$dx^2+dy^2+dz^2-c^2\,dt^2$$

is the element of arc. On taking $i\,ds$ to be the length of the element we have

$$ds^2 = c^2\,dt^2-dx^2-dy^2-dz^2. \tag{1}$$

The reason for choosing $i\,ds$ for the length of the element instead of ds is that in practical problems $(\dot{x}^2+\dot{y}^2+\dot{z}^2)^{\frac{1}{2}}$, the speed of a particle, is always less than c, the speed of light. Hence in (1) ds^2 is positive. The choice of $i\,ds$ instead of ds is merely a matter of convention and has no object other than that of convenience. The sign associated with each space variable is different from that associated with the time variable. This fact is of no great importance mathematically but it is important when we wish to correlate our mathematical results with the phenomena of the real world around us.

Gauss† has shown that an equation between the line element and the coordinate elements, such as (1), characterizes completely the geometry of space. For example, from the equation

$$ds^2 = dx^2+dy^2+dz^2 \tag{2}$$

can be deduced all the well-known properties of three-dimensional Euclidean space. Thus the space-time continuum is completely characterized by (1).

In three-dimensional geometry the trajectory of a particle is generally specified in terms of equations of the form $x = f_1(t)$, $y = f_2(t)$, $z = f_3(t)$, where f_1, f_2, f_3 denote various functional forms. In the space-time continuum defined by (1) these equations specify a curve known as the world line of the particle.

6.10. An approach to relativity dynamics of a particle

Having generalized the relativity formulae our next task is to generalize dynamical theory. In carrying out this task we must bear in mind the great successes of Newtonian theory in dynamics, in astronomy, and in physics generally, so that common sense enjoins us to pursue the following policy. The Newtonian principles of dynamics must be examined in order to discover which, if any, will fit into the framework of Einstein relativity. If none such are found, then those principles are to be used which can be made to fit with the least possible modification.

A dynamical principle can be fitted into the Einstein framework if in its mathematical form no distinction is shown between the properties of the space and time variables, for otherwise the mathematical expressions would not be covariant for the relativity transformations (3), § 6.8. This remark applies to Hamilton's principle (Chap. V), as we proceed to demonstrate.

If T and V denote the kinetic and potential energies respectively and $L = T-V$, then the principle states that $\int\limits_{t_0}^{t_1} L\,dt$ is stationary for a dynamical path. In addition, if the arc of integration does not extend beyond the nearest conjugates

† T. Levi-Civita, *The Absolute Differential Calculus*, chap. v.

of either of the end-points, then the integral is a minimum. Denoting, as usual, a variation by δ the principle can be stated in the form

$$\delta \int_{t_0}^{t_1} L\,dt = 0, \tag{1}$$

where the end-points t_0 and t_1 are fixed.

In our proof of Hamilton's principle the space parameters $q_1, q_2, \ldots, q_n$ were varied as follows. On the extremal we had

$$q_i = s_i(t) \quad (i = 1, 2, \ldots, n), \tag{2}$$

and on a varied curve

$$q_i = s_i(t) + \epsilon_i u_i(t), \tag{3}$$

where ϵ_i is an arbitrary constant and $u_i(t)$ is an arbitrary function of t independent of each ϵ_i. Evidently the time is not varied, so that the theorem established in Chapter V is essentially

$$\int_{t_0}^{t_1} (\delta L)\,dt = 0. \tag{4}$$

Before we try to fit Hamilton's principle into relativity mechanics we must first prove that

$$\int_{t_0}^{t_1} \delta(L\,dt) = 0, \tag{5}$$

where the time is varied as well as the space parameters

$$q_1, q_2, \ldots, q_n,$$

when we pass from a point on the extremal to a point on a neighbouring curve. It is evident that the variation of the q's will lead us once again to Lagrange's equations, (2), § 5.7; we may therefore regard $q_1, q_2, \ldots, q_n$ as constants and vary t only. But a variation in t will give us a variation in $\dot{q}_i$ even if q_i is constant. For

$$\dot{q}_i = \lim_{dt \to 0} \frac{dq_i}{dt}. \tag{6}$$

Therefore if t is varied by the amount δt and the consequent variation in $\dot{q}_i$ is denoted by $\delta\dot{q}_i$, we have

$$\delta\dot{q}_1 = \lim_{dt\to 0}\left\{\frac{dq_1}{d(t+\delta t)} - \frac{dq_1}{dt}\right\} \tag{7}$$

$$= \lim_{dt\to 0}\left\{\frac{dq_1}{dt}\left[1 - \frac{d(\delta t)}{dt}\right] - \frac{dq_1}{dt}\right\} \tag{8}$$

$$= -\dot{q}_1\frac{d(\delta t)}{dt}, \tag{9}$$

where $(\delta t)^2$ and higher powers have been neglected. There are, in addition, corresponding results for $\dot{q}_2, \dot{q}_3,..., \dot{q}_n$.

6.11. Applicability of Hamilton's principle to relativity mechanics†

We now proceed with the proof of equation (5), § 6.10, assuming that the q's remain constant and that the end points of integration are fixed. We have

$$\int_{t_0}^{t_1} \delta(L\,dt) = \int_{t_0}^{t_1} L\,\delta(dt) + \int_{t_0}^{t_1}\left\{\frac{\partial L}{\partial t}\delta t + \sum_{i=1}^{n}\frac{\partial L}{\partial \dot{q}_i}\delta\dot{q}_i\right\}dt \tag{1}$$

$$= \int_{t_0}^{t_1} L\,\delta(dt) + \int_{t_0}^{t_1}\left\{\frac{\partial L}{\partial t}\delta t - \sum_{i=1}^{n}\frac{\partial L}{\partial \dot{q}_i}\dot{q}_i\frac{d(\delta t)}{dt}\right\}dt. \tag{2}$$

on using (9), § 6.10. Integrating by parts and using $\delta t_0 = \delta t_1 = 0$ we have

$$\int_{t_0}^{t_1} \delta(L\,dt) = \int_{t_0}^{t_1}\left\{\frac{\partial L}{\partial t} - \frac{d}{dt}\left(L - \sum_{i=1}^{n}\dot{q}_i\frac{\partial L}{\partial \dot{q}_i}\right)\right\}\delta t dt. \tag{3}$$

But
$$\frac{dL}{dt} = \sum_{i=1}^{n}\frac{\partial L}{\partial q_i}\dot{q}_i + \sum_{i=1}^{n}\frac{\partial L}{\partial \dot{q}_i}\ddot{q}_i + \frac{\partial L}{\partial t}. \tag{4}$$

Hence from Lagrange's equations, (2), § 5.7, we have

$$\frac{dL}{dt} = \sum_{i=1}^{n}\frac{d}{dt}\left(\frac{\partial L}{\partial \dot{q}_i}\right)\dot{q}_i + \sum_{i=1}^{n}\frac{\partial L}{\partial \dot{q}_i}\ddot{q}_i + \frac{\partial L}{\partial t}, \tag{5}$$

† 6.11 and subsequent sections follow the treatment of Levi-Civita, loc. cit., p. 289 et seq. For an alternative treatment see Tolman, loc. cit., §§ 26 and 27.

which is equivalent to

$$\frac{\partial L}{\partial t}-\frac{d}{dt}\left\{L-\sum_{i=1}^{n}\dot{q}_i\frac{\partial L}{\partial \dot{q}_i}\right\} = 0. \tag{6}$$

The proof of (5), § 6.10, is then completed by combining (6) and (3).

6.12. Equations of motion of a particle in relativity mechanics

One of the simplest problems in Newtonian dynamics is that of a particle moving on a smooth surface under no forces other than the normal reaction. Since energy is conserved and there is no potential energy, the velocity must be constant. The principle of least action, §§ 5.6 and 5.19, then states that $\int ds$ is stationary. Thus, the path is a geodesic.

Alternatively we may say that the path is given by

$$\delta\int_{t_0}^{t_1} ds = 0, \tag{1}$$

where
$$ds^2 = dx^2+dy^2+dz^2. \tag{2}$$

The generalization of this result in Einstein mechanics is that the world line of a particle under no forces is a geodesic in the space-time continuum. In other words, the generalization requires the path of a particle under no forces to be specified by

$$\delta\int_{t_0}^{t_1} ds = 0, \tag{3}$$

where
$$ds^2 = c^2\,dt^2-dx^2-dy^2-dz^2. \tag{4}$$

The change in the definition of ds from (2) to (4) is essential since the relativity transformations (3), § 6.8, hold for (4) but not for (2). If v is the velocity of the particle, (3) can be written in the form

$$\delta\int (c^2-v^2)^{\frac{1}{2}}\,dt = 0. \tag{5}$$

For reasons which will appear later there can be no complete analogue of (1) and (2) in Einstein mechanics. But if we are prepared to accept an approximation, a result can be obtained which is not only of great interest in itself but which also explains why such an analogue cannot exist in a simple form. In making the approximation the terms we neglect are of the order v^4/c^4 in comparison with unity, where v is the velocity of the particle and c ($= 3.10^{10}$ cm. per sec.) is the velocity of light. If v is equal to the velocity of the earth in its orbit, then v^4/c^4 is about 10^{-16}.

Terms of order v^2/c^2 must be retained, although in normal motion they are very small. For when v compares with c these terms are of appreciable magnitude and it is then possible to detect differences between Newtonian and Einsteinian mechanics by experimental means.

In Newtonian dynamics, when a particle moves in a conservative field of force, we have

$$\tfrac{1}{2}mv^2+V = \text{constant}, \tag{6}$$

where m is the mass, v the velocity, and V the potential energy. Hence V is of the same order as v^2, and to our degree of approximation we may neglect terms containing V^2/c^4 and higher powers.

In fitting Hamilton's principle into Einsteinian mechanics, a step justified by § 6.11, we may modify the statement of the principle, $\delta \int\limits_{t_0}^{t_1} L\,dt = 0$, in non-essential particulars. This means that the integrand can be multiplied by or increased by a constant, for such modifications have no influence upon the characteristic equations and their solutions. For simplicity we confine ourselves to the case of a particle with unit mass for which

$$L = \tfrac{1}{2}v^2-V. \tag{7}$$

The form the principle then takes for use in relativity is

$$-\delta \int\limits_{t_0}^{t_1} c(c^2-2L)^{\frac{1}{2}}\,dt = 0. \tag{8}$$

For, to our degree of approximation, (i) (8) differs non-essentially from the Newtonian form $\int L\,dt = 0$, and (ii) when the particle is under no forces and $V = 0$, (8) reduces to a form which differs non-essentially from (5).

The world lines of a particle of unit mass in a space-time continuum are then the characteristic equations of (8). Writing

$2L = \dot{x}^2+\dot{y}^2+\dot{z}^2-2V$ and using theorem 7, § 3.2, the world lines are given by

$$\frac{d}{dt}\left\{\frac{c\dot{x}}{(c^2-2L)^{\frac{1}{2}}}\right\} = -\frac{c}{(c^2-2L)^{\frac{1}{2}}}\frac{\partial V}{\partial x} \tag{9}$$

together with two analogous equations for y and z.

6.13. Mass in relativity mechanics

In Newtonian dynamics the equations of motion of a particle, whose mass is m, are given by

$$\frac{d(m\dot{x})}{dt} = -\frac{\partial V}{\partial x}, \tag{1}$$

together with two analogous equations for y and z. Now (9), § 6.12, shows that in relativity dynamics a unit particle with velocity components $(\dot{x}, \dot{y}, \dot{z})$ parallel to the axes has momentum components

$$\left\{\frac{c\dot{x}}{(c^2-2L)^{\frac{1}{2}}}, \frac{c\dot{y}}{(c^2-2L)^{\frac{1}{2}}}, \frac{c\dot{z}}{(c^2-2L)^{\frac{1}{2}}}\right\}. \tag{2}$$

Even if there are no external forces, i.e. $V = 0$, the momentum components become

$$(\beta\dot{x}, \beta\dot{y}, \beta\dot{z}), \tag{3}$$

where $\beta = (1-v^2/c^2)^{-\frac{1}{2}}$, as in § 6.8. Hence if we wish to retain the concept of momentum together with the principle, so fundamental in Newtonian theory, that the rate of change of momentum is proportional to force, we must abandon the principle of constancy of mass. We can retain the principle and account for the factor β in (3) only by introducing the following postulate. A particle which, to an observer at relative rest, appears to have unit mass must appear to have mass β to an observer with relative velocity v. Proportionately a particle which, to an observer at relative rest, appears to have mass m_0 will, to an observer with relative velocity v, appear to have mass m, where

$$m = m_0\left(1-\frac{v^2}{c^2}\right)^{-\frac{1}{2}}. \tag{4}$$

The quantity m_0 is called the invariant or rest mass and m is called the relative mass. According to (4) the relative mass of a particle tends to infinity as its velocity tends to that of light.

Thus in the Einstein theory the velocity of light must be regarded as a limiting velocity which can be approached from below, but never attained by any physical particle.

Equation (4) necessitates a very substantial modification of our physical ideas of matter. But it had already been noticed before the publication of the special theory of relativity, in 1905, that the mass of fast moving electrons could not be constant in the sense required by Newtonian concepts. Accurate experiments, performed from 1901 onwards, have established the truth of (4) and so confirmed Einstein's theory.

In spite of this radical change in the conception of mass, it can be shown that if two particles collide directly their total relative mass and their total momentum is unaltered by the collision.†

6.14. Energy in relativity mechanics

The far-reaching changes in our views of matter required by the preceding results must necessarily be accompanied by commensurate changes in our views of energy.

We must first find an acceptable definition of energy for use in relativity mechanics and in this search we must be guided once more by the principles outlined at the beginning of § 6.10. Since Hamilton's principle is equally applicable to both Newton's and Einstein's theories (§ 6.10), a suitable definition of energy can be based upon the Lagrangian function L $(= T-V)$. When the forces are conservative the kinetic energy T is a homogeneous function of degree two in the variables $\dot{q}_1, \dot{q}_2,..., \dot{q}_n$, as shown in § 5.14. Since the potential energy V is independent of these variables it follows that

$$\sum_{i=1}^{n} \frac{\partial L}{\partial \dot{q}_i} \dot{q}_i - L = 2T - L = T+V, \tag{1}$$

and therefore, in the Newtonian case, the left-hand side of (1) is equal to the total energy.

Before a definition can be accepted for use with the Einstein theory its mathematical expression must remain covariant for the relativity transformations (3), § 6.8, and distinctions between

† A. S. Eddington, loc. cit., § 12; R. C. Tolman, loc. cit., § 23.

the space and time variables eliminated. These requirements are not fulfilled by the left-hand side of (1). But if we are prepared to accept an approximation, as in § 6.12 (i.e. we retain terms of order v^2/c^2 and neglect terms of order v^4/c^4 and smaller), further progress can be made. The arguments of § 6.12 then show that the requirements will be fulfilled if L is replaced by $-c(c^2-2L)^{\frac{1}{2}}$ and that the left-hand side of (1) can serve satisfactorily as the definition of energy in the Einstein theory.

For a particle of unit mass we have

$$-c(c^2-2L)^{\frac{1}{2}} = -c(c^2-\dot{x}^2-\dot{y}^2-\dot{z}^2+2V)^{\frac{1}{2}}. \tag{2}$$

To our degree of approximation, with $q_1 = x$, $q_2 = y$, $q_3 = z$, the energy of a unit particle is

$$\sum_{i=1}^{n} \frac{\partial\{-c(c^2-2L)^{\frac{1}{2}}\}}{\partial \dot{q}_i}\dot{q}_i + c(c^2-2L)^{\frac{1}{2}} = c^2+\tfrac{1}{2}v^2+V. \tag{3}$$

Thus even when at rest and not acted on by any forces, i.e. $v = 0$ and $V = 0$, the particle possesses a store of energy equal to c^2 ergs ($= 9.10^{20}$ ergs).

For a particle of relative mass m the store of energy E is given by

$$E = mc^2. \tag{4}$$

Evidently this cannot be classified in either of the two traditional categories of energy, kinetic and potential, and it must therefore represent energy in a new form. Enormous quantities of energy other than kinetic and potential occur in nature associated with the phenomena of radioactivity and atomic fission. In both these phenomena quantities of energy of the order mc^2 are released and after the release of the energy it is found that the total amount of matter has diminished. The matter lost has been transformed into energy in accordance with (4), an equation fundamental in modern investigations in nuclear physics.

6.15. Further observations

Since Hamilton's principle can be applied to relativity mechanics it follows that the calculus of variations plays an important part in the development of relativity theory. The theory has been extended so as to include all gravitational and electromagnetic phenomena and in these extensions the new principles have been put into simple and elegant forms by means of variational methods. For an account of these developments the reader is referred to the standard works on relativity.†

† A. S. Eddington, loc. cit., p. 139 et seq. In particular the definition and use of the Hamiltonian operator. H. Weyl, loc. cit., §§ 28 and 36. R. C. Tolman, loc. cit., § 87.

CHAPTER VII

APPROXIMATION METHODS WITH APPLICATIONS TO PROBLEMS OF ELASTICITY

7.1. Introduction

OUR account of the Calculus of Variations has so far been based upon the properties of Euler's differential equation (8), § 1.4, and the generalizations of this equation given in Chapters III and IV. A new and fruitful line of development was inaugurated by Rayleigh and later used by Ritz and others to deal with problems of elasticity and numerous other branches of Applied Mathematics, including Quantum theory. In this chapter, after a brief introduction to the mathematical theory of elasticity, we shall develop the Rayleigh–Ritz method and then illustrate its use in practice by an application to the Saint–Venant torsion problem.

We commence by an investigation in which the stationary value of an integral can be found either by the use of Euler's equation or by the use of the Rayleigh–Ritz method. The two procedures can then be contrasted with each other.

7.2. Illustration using Euler's equation

Consider the problem of finding the functional form of y which renders the integral

$$I = \int_{-1}^{1} (1-x^2)\left(\frac{dy}{dx}\right)^2 dx \tag{1}$$

stationary subject to the condition

$$1 = \int_{-1}^{1} y^2\, dx. \tag{2}$$

This type of problem is solved in Chapter IV and its Eulerian equation [see (12), § 4.2] is

$$\frac{\partial}{\partial y}\{(1-x^2)y_1^2-\lambda y^2\}-\frac{d}{dx}\left[\frac{\partial}{\partial y_1}\{(1-x^2)y_1^2-\lambda y^2\}\right] = 0, \tag{3}$$

where λ is an undetermined multiplier and $y_1 = dy/dx$. This simplifies to

$$\frac{d}{dx}\left\{(1-x^2)\frac{dy}{dx}\right\}+\lambda y = 0, \tag{4}$$

which is the well-known Legendre equation. If we write

$$\lambda = n(n+1)$$

the two solutions of (4) are usually denoted by $P_n(x)$ and $Q_n(x)$, the first and second Legendre function respectively.† The general solution of (4) is

$$y = AP_n(x)+BQ_n(x), \tag{5}$$

where A and B are arbitrary constants. If n is a positive integer $P_n(x)$ is a polynomial which is the coefficient of h^n in the expansion of $(1-2xh+h^2)^{-\frac{1}{2}}$ in positive integral powers of h. The first six of these polynomials are given in the following table:

n	λ	$P_n(x)$
0	0	1
1	2	x
2	6	$\frac{1}{2}(3x^2-1)$
3	12	$\frac{1}{2}(5x^3-3x)$
4	20	$\frac{1}{8}(35x^4-30x^2+3)$
5	30	$\frac{1}{8}(63x^5-70x^3+15x)$

It is easily verified that

$$\left(\frac{2n+1}{2}\right)^{\frac{1}{2}} P_n(x)$$

satisfies (2) when $n = 0, 1, 2, 3, 4$, or 5. In fact it satisfies (2) for all positive integral values of n.

The introduction of condition (2) enlarges considerably the range of functions which render (1) stationary. For the omission of (2) is equivalent to writing $\lambda = 0$ in (3), and then from (5) the most general function which renders (1) stationary is

$$y = AP_0(x)+BQ_0(x) = A+B\log\left(\frac{1+x}{1-x}\right), \tag{6}$$

where A and B are arbitrary constants.

† Whittaker and Watson, *Modern Analysis*, chap. xv, where a comprehensive account of these functions can be found.

7.3. Illustration using the Rayleigh–Ritz method

The object of the Rayleigh–Ritz method is to replace the problem of finding the maxima and minima of integrals by that of finding the maxima and minima of functions of several variables. This is soluble by the ordinary processes of the differential and integral calculus.† A simplified account of the method is as follows.

First it is assumed that y, the dependent variable, can be expressed in terms of known functions of x in a form which involves unknown coefficients or parameters. On substituting for y in the integral (1), § 7.2, the value of I can be found in terms of these parameters. The theory of maximum and minimum values of functions of several variables then enables us to determine the values of the parameters which render I stationary.

Let us assume that the integral I, (1), § 7.2, is stationary when

$$y = a+bx+cx^2, \tag{1}$$

where a, b, c remain to be determined. Substituting for y we have

$$I = \tfrac{4}{3}(b^2+\tfrac{4}{5}c^2). \tag{2}$$

From (2), § 7.2, we have

$$0 = -1+2\left(a^2+\frac{b^2}{3}+\frac{2ac}{3}+\frac{c^2}{5}\right). \tag{3}$$

Denoting the right-hand side of (3) by ϕ, the values of a, b, c for which I is stationary are given by‡

$$\frac{\partial I/\partial a}{\partial\phi/\partial a} = \frac{\partial I/\partial b}{\partial\phi/\partial b} = \frac{\partial I/\partial c}{\partial\phi/\partial c} = \mu. \tag{4}$$

Evaluation of (4) gives us

$$\frac{0}{4(a+c/3)} = \frac{8b/3}{4b/3} = \frac{32c/15}{4(a/3+c/5)} = \mu, \tag{5}$$

and the possible solutions are as follows:

(i) $b = c = \mu = 0$; (ii) $a = c = 0,\ \mu = 2$;

(iii) $c = -3a,\ b = 0,\ \mu = 6$.

Using (3) we have in addition for case (i) $a = 1/\surd 2$; for case

† Courant, *Differential and Integral Calculus*, vol. ii, p. 183.

‡ Courant, loc. cit., vol. ii, p. 194.

(ii) $b = \sqrt{(3/2)}$, and for case (iii) $a = \sqrt{(5/8)}$. The final results are then

μ	y
0	$1/\sqrt{2}$
2	$\sqrt{\frac{3}{2}}x$
6	$\sqrt{\frac{5}{8}}(1-3x^2)$

On comparison with the table in § 7.2 these functions are evidently equal to $\left(\dfrac{2n+1}{2}\right)^{\frac{1}{2}} P_n(x)$ for the three cases $n = 0, 1, 2$. They satisfy both the Eulerian equation (4), § 7.2, with $\mu = \lambda$, and condition (2), § 7.2. They give us therefore an exact solution of the problem of minimizing the integral I, (1), § 7.2, subject to condition (2), § 7.2.

EXAMPLE. Obtain similarly solutions proportional to $P_3(x)$, $P_4(x)$, and $P_5(x)$ (assume $y = a+bx+cx^2+dx^3$, etc.).

7.4. Rayleigh's method

In the example above an exact solution has been obtained, but in most cases the Rayleigh–Ritz method gives approximations only. To illustrate the degree of accuracy which can be attained we shall consider one of Rayleigh's† own examples, the problem of vibrating strings. We first obtain the exact solution and then compare it with the result obtained by Rayleigh's method.

Consider the transverse vibrations of a taut uniform string whose end-points are fixed. The tension is τ, the length l, the coordinates of a point on it (x, y) and those of the end points $(\pm\frac{1}{2}l, 0)$. If σ is the mass per unit length, then the kinetic energy of an element of length ds is $\frac{1}{2}\sigma\, ds(\partial y/\partial t)^2$. Since the string is taut $\partial y/\partial x$ is small compared with unity, so that

$$ds = (dx^2+dy^2)^{\frac{1}{2}} = dx,$$

on neglecting infinitesimals of the third order. Hence the total kinetic energy of the string T is given by the equation

$$T = \tfrac{1}{2}\sigma \int\limits_{-\frac{1}{2}l}^{\frac{1}{2}l} \left(\frac{\partial y}{\partial t}\right)^2 dx. \tag{1}$$

† Lord Rayleigh, *Theory of Sound*, vol. i, § 89.

Since the change in length of the string during the vibration is negligible we may take the tension τ to be constant. The work done in extending an element from dx to ds is

$$\tau(ds-dx) = \tau\{(dx^2+dy^2)^{\frac{1}{2}}-dx\} = \tfrac{1}{2}\tau\left(\frac{\partial y}{\partial x}\right)^2 dx,$$

neglecting $(\partial y/\partial x)^4\,dx$ and smaller terms. Hence the potential energy of the string V is given by

$$V = \tfrac{1}{2}\tau \int\limits_{-\frac{1}{2}l}^{\frac{1}{2}l} \left(\frac{\partial y}{\partial x}\right)^2 dx. \tag{2}$$

By Hamilton's principle, (5), § 5.5, it follows that

$$\int\limits_{t_0}^{t_1} (T-V)\,dt = \frac{1}{2}\int\limits_{t_0}^{t_1}\int\limits_{-\frac{1}{2}l}^{\frac{1}{2}l} \left\{\sigma\left(\frac{\partial y}{\partial t}\right)^2 - \tau\left(\frac{\partial y}{\partial x}\right)^2\right\} dxdt \tag{3}$$

is stationary. The characteristic equation for this integral, given by (6), § 3.8, is

$$\sigma\frac{\partial^2 y}{\partial t^2} = \tau\frac{\partial^2 y}{\partial x^2}. \tag{4}$$

The most general solution is

$$y = f_1\left(x-t\sqrt{\frac{\tau}{\sigma}}\right)+f_2\left(x+t\sqrt{\frac{\tau}{\sigma}}\right), \tag{5}$$

where f_1 and f_2 are arbitrary functions. In the slowest or fundamental mode the string vibrates as a whole without nodes except at the fixed end points. The particular form of (5) which satisfies these end conditions is

$$y = a\cos\left(\frac{\pi x}{l}\right)\cos\left\{\frac{\pi t}{l}\sqrt{\left(\frac{\tau}{\sigma}\right)}+\epsilon\right\}, \tag{6}$$

where a and ϵ are constants. Hence the period of the fundamental mode is

$$2l\sqrt{\frac{\sigma}{\tau}} \tag{7}$$

and the exact solution of the problem is obtained.

The Rayleigh method assumes that the vibrations of the string are periodic but does not attempt to find the exact form. Instead, from the data, a simple relationship is inferred which must be approximate to the exact form of vibration. In this case we can account most simply for (i) the period $2\pi/p$ by the factor $\cos(pt+\epsilon)$, and (ii) the end conditions, $y = 0$, $x = \pm\frac{1}{2}l$ for all values of t, by the factor $\{1-(4x^2/l^2)\}$. We therefore assume that

$$y = a\cos(pt+\epsilon)\left(1-\frac{4x^2}{l^2}\right), \tag{8}$$

is an approximation to the mathematical expression for the vibration. Here a and ϵ are constants, and p is to be determined as below. The form of vibration assumed is parabolic.

Rayleigh continues by using the following principle:† when a conservative dynamical system, with a finite number of degrees of freedom, vibrates in a normal mode, then the periods are stationary. From which it follows that the mean kinetic and potential energies taken over a period are equal. In the case of a string, for which there are an infinite number of periods of vibration, the principle is still true for the fundamental mode.

From (8) and (1) we have

$$T = \tfrac{4}{15}\sigma l a^2 p^2 \sin^2(pt+\epsilon), \tag{9}$$

and from (8) and (2) we have

$$V = \frac{8\tau a^2}{3l}\cos^2(pt+\epsilon). \tag{10}$$

On averaging these expressions for T and V over the time period $2\pi/p$ and equating the results we get

$$\frac{2\pi}{p} = \frac{\pi}{\sqrt{10}}\,2l\sqrt{\frac{\sigma}{\tau}}. \tag{11}$$

Comparison between (11) and the exact result (7) shows that the approximation is slightly too small in the ratio $\pi/\sqrt{10}$ ($= 0{\cdot}9936$). Thus in spite of the somewhat crude nature of assumption (8), the final result differs from the correct result by less than 0·7 per cent., a degree of accuracy sufficiently high for most practical purposes. We also observe that this result is arrived at by relatively short and simple calculations.

† Rayleigh, loc. cit., vol. i, § 88. This is sometimes known as Rayleigh's principle and is proved in most books on dynamics, e.g. Whittaker, *Analytical Dynamics*, § 82; Ramsey, *Dynamics*, Part II, § 10.6.

7.5. The Rayleigh–Ritz method

We first consider integrals of the type studied in § 3.5, namely

$$I = \int_a^b F(x, y, y_1, y_2,..., y_n)\,dx, \tag{1}$$

where a and b are given, F is a known function and $y_m = d^m y/dx^m$. In order to determine the functional form of y which renders (1) stationary we assume that y can be expressed in the form

$$y = \sum_{i=1}^{n} a_i f_i(x), \tag{2}$$

where the functions $f_i(x)$ $(i = 1, 2,..., n)$ are arbitrarily chosen and the parameters a_i $(i = 1, 2,..., n)$ are for the moment undetermined. On substituting for y in (1) and evaluating the integral we obtain an expression for I in terms of the parameters a_i $(i = 1, 2,..., n)$. By the usual methods of the differential calculus I is stationary if the parameters a_i are chosen so that†

$$\frac{\partial I}{\partial a_i} = 0 \quad (i = 1, 2,..., n). \tag{3}$$

On solving these simultaneous equations for a_i and substituting in (2) the form of y is determined.‡

The result obtained may be the exact answer, as in § 7.3, but in most cases it will be an approximation only. Its closeness to the exact answer will depend very largely upon the initial choice of the functions $f_i(x)$ and can be tested by substituting for y in the Eulerian equation for the integral [(12), § 3.5]. If the left side vanishes, then the answer is exact, but if it differs from zero then the answer is approximate only. In some cases this difference from zero is used as a measure of the degree of approximation attained.

As most applications of the Rayleigh–Ritz method lead to approximate answers, the problem of estimating the degree of approximation is of great importance. In some elasticity investigations, as we shall show later, the exact answer can be enclosed between narrow bounds by a series of approximations from above and below. In investigations for which such bounds

† Courant, loc. cit., vol. ii, p. 194.

‡ Ritz, *Journal für reine und angewandte Mathematik*, **125** (1909), 1–61.

cannot be obtained the problem of estimating the degree of approximation can become very difficult.

Care must be exercised in choosing the functions $f_i(x)$. For upon this choice depends both the degree of approximation attained and the amount of computation involved in obtaining the answer. The nature of the problem must be carefully studied and used as a guide. For example if there is any symmetry in the problem this should be reflected in the choice of $f_i(x)$. If the exact answer is required to satisfy certain prescribed conditions, then it is advisable to choose the functions $f_i(x)$ so that each one satisfies these conditions, if possible. An illustration of this point is given in the problem of vibrating strings dealt with in § 7.4 where for all values of the time t, $y = 0$ when $x = \pm\frac{1}{2}l$. The form assumed for y therefore contained the factor $\{1-(4x^2/l^2)\}$.

If the degree of approximation is insufficient, closer approximations can be obtained by increasing the value of n in (2). In some cases $\sum_{i=1}^{n} a_i f_i(x)$ converges to the exact answer as n tends to infinity, but there is no certainty that this will always happen. If, in order to obtain close approximations, a large number of functions $f_i(x)$ are utilized, it is advisable to choose a sequence of functions which are complete. The functions $f_i(x)$ $(i = 1, 2,\ldots)$, are said to be complete if it is possible to express an arbitrary function in the form $\sum_{i=1}^{\infty} a_i f_i(x)$.

A more precise definition of completeness is as follows. A function of x is said to be piece-by-piece continuous in an interval (or piecewise continuous) if (i) on dividing the interval into a finite number of sub-intervals it is continuous in each sub-interval, and (ii) it tends to finite limits as x approaches any of the sub-interval boundaries. If for any piece-by-piece continuous function $F(x)$ a set of coefficients a_i $(i = 1, 2,\ldots)$ can be found such that

$$\lim_{n\to\infty} \int_a^b \Big\{F(x)-\sum_{i=1}^{n} a_i f_i(x)\Big\}^2\, dx = 0, \tag{4}$$

then the functions $f_i(x)$ are said to be complete.

The trigonometric functions $\sin nx$, $\cos nx$ $(n = 1, 2,...)$, and also the Legendre polynomials $P_i(x)$ $(i = 1, 2,...)$, possess this property. A very large class of functions which are complete in the above sense are the Sturm–Liouville functions.

7.6. Sturm–Liouville functions

These functions are solutions of the differential equation

$$\frac{d}{dx}\left(p\frac{dy}{dx}\right)+(\lambda q-r)y = 0, \tag{1}$$

where p, q, r are continuous functions of x in a given interval (a, b) and λ is a parameter independent of x. Most of the well-known differential equations of mathematical physics can be obtained from (1) by suitable choice of p, q, and r. For example $p = q = 1$, $r = 0$ leads to the differential equation of simply periodic function; $p = 1-x^2$, $q = 1$, $r = 0$ leads to the Legendre equation of § 7.2; $p = q = x$, $r = -n^2/x$ leads to the Bessel equation. Other special cases of (1) are the Mathieu equation, Gauss's hypergeometric equation, the equation for the Hermite, Tschebyschef, and Laguerre polynomials and many of their associated functions. In spite of their great diversity these functions have many properties in common.

In most practical applications the solutions of (1) are required either to satisfy conditions at the ends of the interval (a, b), known as boundary conditions, or to satisfy conditions of convergence. In such cases the parameter λ cannot assume any value but is restricted to take one of a determinate sequence of values $\lambda_1, \lambda_2, \lambda_3,..., \lambda_n$. These numbers are known as eigenvalues and to each eigenvalue there corresponds a solution of (1) which is called an eigenfunction.

As an illustration consider the problem of the vibrating string dealt with in § 7.4. On writing $y = u(x)\cos\alpha t$ in (4), § 7.4 we obtain

$$\tau\frac{d^2u}{dx^2}+\sigma\alpha^2 u = 0, \tag{2}$$

where, for brevity, u has replaced $u(x)$. This equation is of the Sturm–Liouville form with $p = q = 1$, $r = 0$, $\lambda = \sigma\alpha^2/\tau$, and the solution is

$$u(x) = A\cos\lambda^{\frac{1}{2}}x+B\sin\lambda^{\frac{1}{2}}x. \tag{3}$$

If for all values of the time t we must have $y = 0$ at the end points $x = \pm\frac{1}{2}l$, then $A = 0$ and $\lambda = 4n^2\pi^2/l$, where n is an integer. This is the nth eigenvalue and the corresponding eigenfunction is given by

$$u_n(x) = B\sin\left(\frac{2n\pi x}{l}\right). \tag{4}$$

The Legendre equation (4), § 7.2, has important applications to problems of quantum theory† in which conditions of convergence can be satisfied only if $\lambda = n(n+1)$, where n is a positive integer. This, then, is the nth eigenvalue and the Legendre polynomial $P_n(x)$ is accordingly the nth eigenfunction for such problems.

In the case where $q > 0$ it can be proved that there are an enumerable infinity‡ of eigenvalues and eigenfunctions and that these functions have the property of completeness. It can be shown that any continuous function can be developed in terms of Sturm–Liouville functions and that such development converges or diverges in the same way as the cosine development of the function.

Proofs of this statement and comprehensive accounts of the properties of these functions are to be found in numerous works on differential equations and quantum theory.§

7.7. The case of several independent variables

Integrals of the type

$$I = \iint F(x, y, z; p, q)\, dxdy, \tag{1}$$

where x and y are independent variables, z is a dependent variable, $p = \partial z/\partial x$, and $q = \partial z/\partial y$, have already been studied in § 3.6. The determination of stationary values of such integrals can be approximately effected by Rayleigh–Ritz methods and there is no difficulty in extending the process to integrals with several independent variables.

† E. C. Kemble, *The Fundamental Principles of Quantum Mechanics*, § 28.

‡ An enumerable infinity of numbers can be put into one-one correspondence with the integers.

§ E. Lindsay Ince, *Ordinary Differential Equations*, chap. xi, in particular p. 276 et seq.; E. C. Titchmarsh, *Eigenfunction Expansions*; Courant and Hilbert, *Methoden der mathematischen Physik*, vol. i, chap. vi; E. C. Kemble, loc. cit., § 23.

For (1) we assume an expansion of the form

$$z = \sum_{i=1}^{n} a_i f_i(x, y), \tag{2}$$

where $f_i(x,y)$ are suitably chosen functions and a_i $(i = 1, 2,..., n)$ are for the moment undetermined parameters. On substituting for z in (1) I can be evaluated as a function of the parameters a_i and is stationary if these parameters are chosen to satisfy the n simultaneous equations

$$\frac{\partial I}{\partial a_i} = 0 \quad (i = 1, 2,..., n). \tag{3}$$

The amount of computation involved and also the degree of approximation attained both depend upon the choice of the functions $f_i(x,y)$. If, for example, z is symmetrical in x and y, this should be reflected in the choice of $f_i(x,y)$. Again, if z satisfies certain boundary conditions it is advisable that each function $f_i(x,y)$ should also satisfy these conditions, if possible. If z satisfies Laplace's equation in two dimensions then, in general, each function $f_i(x,y)$ should also satisfy Laplace's equation. In this case we may sometimes choose $f_i(x,y)$ from the real or imaginary polynomial parts of $(x+iy)^n$, where $i = \sqrt{(-1)}$ and n is a positive integer.

Since the development of Rayleigh–Ritz methods has largely centred round elasticity problems, we shall illustrate our remarks by applications to the theory of elasticity and particularly to the Saint-Venant problem of torsion in prisms. To acquire a better understanding of the nature of this problem we shall, in the next three sections, give a brief account of the principles underlying the theory of elasticity.

7.8. The specification of strain

The simplest examples of elasticity are those associated with the extension or compression of thin rods. If a rod is extended or strained by an amount x beyond its natural length a it exerts a tension $\lambda x/a$ (Hooke's law), where λ is known as Young's modulus. The potential energy of the extension is $\lambda x^2/2a$. These results are true only if x does not exceed a quantity known as the proportional limit, for when this is exceeded Hooke's law no longer applies.† We shall restrict ourselves entirely to the case when this limit is not exceeded.

As a concrete illustration: for certain types of steel the proportional limit is reached when under a load of 10 tons weight per square inch. If the extension is defined as the change in length per unit length, then under a load of 6 tons weight the extension of steel is 0·00046.

For the case of an elastic body in the shape of a plane lamina let (x, y) and $(x+\xi, y+\eta)$ be the coordinates of two neighbouring points P and P_1 respectively when the body is unstrained. When the body is strained let P be displaced to Q, whose coordinates are $(x+u, y+v)$, where u and v are functions of x and y. Then P_1 will be displaced to Q_1, whose coordinates are

$$\left[\left(x+\xi+u+\frac{\partial u}{\partial x}\xi+\frac{\partial u}{\partial y}\eta\right),\quad \left(y+\eta+v+\frac{\partial v}{\partial x}\xi+\frac{\partial v}{\partial y}\eta\right)\right],$$

where terms of the order ξ^2, η^2, $\xi\eta$ and smaller have been neglected. Thus before the strain the projections of PP_1 on the axes are (ξ, η) and after the strain the projections of QQ_1 on the axes are

$$\left[\left(\xi+\frac{\partial u}{\partial x}\xi+\frac{\partial u}{\partial y}\eta\right),\quad \left(\eta+\frac{\partial v}{\partial x}\xi+\frac{\partial v}{\partial y}\eta\right)\right].$$

These changes in the projections are due partly to rotation and partly to deformation or strain. The rotational part causes no strain and so is of no interest to us. The terms due to deformation must therefore be disentangled from those due to rotation, and this is done by means of the following two identities:

$$\frac{\partial u}{\partial x}\xi+\frac{\partial u}{\partial y}\eta \equiv \frac{\partial u}{\partial x}\xi+\frac{1}{2}\left(\frac{\partial u}{\partial y}+\frac{\partial v}{\partial x}\right)\eta-\frac{1}{2}\left(\frac{\partial v}{\partial x}-\frac{\partial u}{\partial y}\right)\eta, \tag{1}$$

$$\frac{\partial v}{\partial x}\xi+\frac{\partial v}{\partial y}\eta \equiv \frac{1}{2}\left(\frac{\partial v}{\partial x}+\frac{\partial u}{\partial y}\right)\xi+\frac{\partial v}{\partial y}\eta+\frac{1}{2}\left(\frac{\partial v}{\partial x}-\frac{\partial u}{\partial y}\right)\xi. \tag{2}$$

† When x exceeds a quantity known as the elastic limit the material acquires a permanent deformation. In many materials the difference between the proportional and elastic limits is small and frequently the distinction between these two limits is disregarded. For steel the elastic limit is reached under a stress of 16 tons per square inch.

For a sufficiently small strain the terms

$$-\frac{1}{2}\left(\frac{\partial v}{\partial x}-\frac{\partial u}{\partial y}\right)\eta \quad \text{and} \quad +\frac{1}{2}\left(\frac{\partial v}{\partial x}-\frac{\partial u}{\partial y}\right)\xi$$

correspond to a rotation† through the small angle $\frac{1}{2}\left(\frac{\partial v}{\partial x}-\frac{\partial u}{\partial y}\right)$. Therefore the terms which correspond to pure strain are given by

$$\frac{\partial u}{\partial x}\xi+\frac{1}{2}\left(\frac{\partial u}{\partial y}+\frac{\partial v}{\partial x}\right)\eta, \tag{3}$$

$$\frac{1}{2}\left(\frac{\partial v}{\partial x}+\frac{\partial u}{\partial y}\right)\xi+\frac{\partial u}{\partial y}\eta. \tag{4}$$

These ideas can be extended to three dimensions. Let (x, y, z) and $(x+\xi, y+\eta, z+\zeta)$ be respectively the coordinates of P and P_1, two neighbouring points in an elastic body, and let P be displaced to Q, whose coordinates are $(x+u, y+v, z+w)$, where u, v, and w are functions of x, y, and z. Then P_1 will be displaced to Q_1, whose coordinates are

$$\left\{\left(x+\xi+u+\frac{\partial u}{\partial x}\xi+\frac{\partial u}{\partial y}\eta+\frac{\partial u}{\partial z}\zeta\right), \quad \left(y+\eta+v+\frac{\partial v}{\partial x}\xi+\frac{\partial v}{\partial y}\eta+\frac{\partial v}{\partial z}\zeta\right), \quad \left(z+\zeta+w+\frac{\partial w}{\partial x}\xi+\frac{\partial w}{\partial y}\eta+\frac{\partial w}{\partial z}\zeta\right)\right\}.$$

The differences in the projections of QQ_1 and PP_1 on the coordinate axes can be written as follows:

$$\left.\begin{aligned}
\frac{\partial u}{\partial x}\xi+\frac{\partial u}{\partial y}\eta+\frac{\partial u}{\partial z}\zeta &\equiv \frac{\partial u}{\partial x}\xi+\frac{1}{2}\left(\frac{\partial u}{\partial y}+\frac{\partial v}{\partial x}\right)\eta+ \\
&\quad +\frac{1}{2}\left(\frac{\partial u}{\partial z}+\frac{\partial w}{\partial x}\right)\zeta-\frac{1}{2}\left(\frac{\partial v}{\partial x}-\frac{\partial u}{\partial y}\right)\eta+\frac{1}{2}\left(\frac{\partial u}{\partial z}-\frac{\partial w}{\partial x}\right)\zeta, \\
\frac{\partial v}{\partial x}\xi+\frac{\partial v}{\partial y}\eta+\frac{\partial v}{\partial z}\zeta &\equiv \frac{1}{2}\left(\frac{\partial v}{\partial x}+\frac{\partial u}{\partial y}\right)\xi+\frac{\partial v}{\partial y}\eta+ \\
&\quad +\frac{1}{2}\left(\frac{\partial w}{\partial y}+\frac{\partial v}{\partial z}\right)\zeta+\frac{1}{2}\left(\frac{\partial v}{\partial x}-\frac{\partial u}{\partial y}\right)\xi-\frac{1}{2}\left(\frac{\partial w}{\partial y}-\frac{\partial v}{\partial z}\right)\zeta, \\
\frac{\partial w}{\partial x}\xi+\frac{\partial w}{\partial y}\eta+\frac{\partial w}{\partial z}\zeta &\equiv \frac{1}{2}\left(\frac{\partial w}{\partial x}+\frac{\partial u}{\partial z}\right)\xi+\frac{1}{2}\left(\frac{\partial w}{\partial y}+\frac{\partial v}{\partial z}\right)\eta+ \\
&\quad +\frac{\partial w}{\partial z}\zeta-\frac{1}{2}\left(\frac{\partial u}{\partial z}-\frac{\partial w}{\partial x}\right)\xi+\frac{1}{2}\left(\frac{\partial w}{\partial y}-\frac{\partial v}{\partial z}\right)\eta.
\end{aligned}\right\} \tag{5}$$

† A. S. Ramsey, *Dynamics*, vol. ii, p. 61.

It can be seen that the terms

$$\left.\begin{aligned}
&-\frac{1}{2}\left(\frac{\partial v}{\partial x}-\frac{\partial u}{\partial y}\right)\eta+\frac{1}{2}\left(\frac{\partial u}{\partial z}-\frac{\partial w}{\partial x}\right)\zeta,\\
&\frac{1}{2}\left(\frac{\partial v}{\partial x}-\frac{\partial u}{\partial y}\right)\xi-\frac{1}{2}\left(\frac{\partial w}{\partial y}-\frac{\partial v}{\partial z}\right)\zeta,\\
&-\frac{1}{2}\left(\frac{\partial u}{\partial z}-\frac{\partial w}{\partial x}\right)\xi+\frac{1}{2}\left(\frac{\partial w}{\partial y}-\frac{\partial v}{\partial z}\right)\eta
\end{aligned}\right\} \tag{6}$$

correspond to small rotations through the angles

$$\frac{1}{2}\left(\frac{\partial v}{\partial x}-\frac{\partial u}{\partial y}\right),\quad \frac{1}{2}\left(\frac{\partial w}{\partial y}-\frac{\partial v}{\partial z}\right),\quad \frac{1}{2}\left(\frac{\partial u}{\partial z}-\frac{\partial w}{\partial x}\right)$$

about the x, y, z axes respectively.†

It therefore follows that the part of (5) which corresponds to a pure strain of the elastic body is given by

$$\left.\begin{aligned}
&\frac{\partial u}{\partial x}\xi+\frac{1}{2}\left(\frac{\partial u}{\partial y}+\frac{\partial v}{\partial x}\right)\eta+\frac{1}{2}\left(\frac{\partial u}{\partial z}+\frac{\partial w}{\partial x}\right)\zeta,\\
&\frac{1}{2}\left(\frac{\partial v}{\partial x}+\frac{\partial u}{\partial y}\right)\xi+\frac{\partial v}{\partial y}\eta+\frac{1}{2}\left(\frac{\partial w}{\partial y}+\frac{\partial v}{\partial z}\right)\zeta,\\
&\frac{1}{2}\left(\frac{\partial w}{\partial x}+\frac{\partial u}{\partial z}\right)\xi+\frac{1}{2}\left(\frac{\partial w}{\partial y}+\frac{\partial v}{\partial z}\right)\eta+\frac{\partial w}{\partial z}\zeta.
\end{aligned}\right\} \tag{7}$$

There are only six independent coefficients of ξ, η, ζ in these expressions. They are known as the components of strain and are not all independent but are related by equations known as conditions of compatibility.

As an illustration of these formulae we consider the Saint-Venant torsion problem. A prism of elastic isotropic material is bounded by a cylinder and two planes perpendicular to the generators of the cylinder. Couples act in the plane ends so that the prism is twisted about an axis perpendicular to them. Taking this axis as the z-axis let the point P, whose coordinates in the unstrained state are (x, y, z) be displaced by the strain to P_1, whose coordinates are $(x+u, y+v, z+w)$. Then, for reasons

† A. S. Ramsey, *Dynamics*, vol. ii, p. 66; E. T. Whittaker, *Analytical Dynamics*, p. 17.

which will appear later (§ 7.12 below), it may be assumed that

$$u = -\tau yz; \quad v = \tau zx; \quad w = \tau\phi(x,y). \tag{8}$$

Here the constant τ is the angular measure of the twist per unit length of prism and $\phi(x,y)$ is a function of x and y only—for brevity we shall write it ϕ. The expressions (7) then become

$$\left.\begin{aligned} &\tfrac{1}{2}\tau\left(\frac{\partial\phi}{\partial x}-y\right)\zeta \\ &\tfrac{1}{2}\tau\left(\frac{\partial\phi}{\partial y}+x\right)\zeta \\ \tfrac{1}{2}\tau\left(\frac{\partial\phi}{\partial x}-y\right)\xi+&\tfrac{1}{2}\tau\left(\frac{\partial\phi}{\partial y}+x\right)\eta \end{aligned}\right\}. \tag{9}$$

7.9. The specification of stress

The forces acting on a small element of a continuous elastic medium are of two kinds. The first are body forces such as gravity which are proportional to the volume and, in most cases, to the density of the element. The second are forces which arise from the actions and reactions between the surface of the element and the surface of the surrounding medium in contact with it; they are proportional to the areas of contact. It is assumed that the ratio of the second kind of force to its area of contact tends to a limit as the magnitude of the area tends to zero. This limit, which is a force per unit area, is known as the stress.

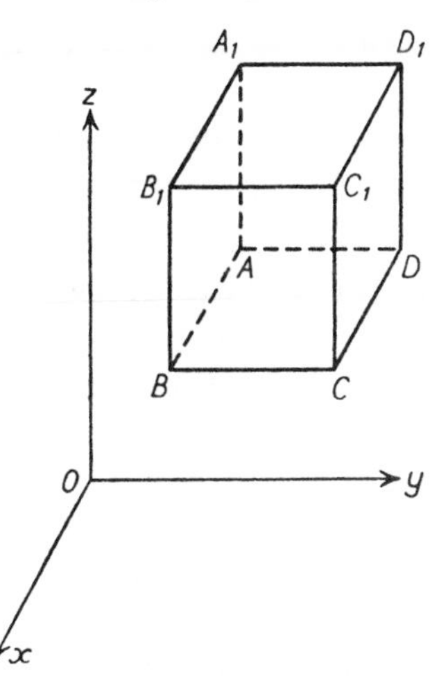

FIG. VII. 1.

Consider a small element of the medium in the form of a rectangular parallelepiped $ABCD$, $A_1\,B_1\,C_1\,D_1$, Fig. VII. 1, and let $AB = dx$, $AD = dy$, $AA_1 = dz$. Let the stress components of the action between this element and the surrounding medium across the area ADD_1A_1 be (X_x, X_y, X_z) parallel to the axes. It will be assumed that these stresses act at the centre of the area ADD_1A_1. Across the parallel area BCC_1B_1 the stress components are

$$\left(X_x+\frac{\partial X_x}{\partial x}\,dx,\ X_y+\frac{\partial X_y}{\partial x}\,dx,\ X_z+\frac{\partial X_z}{\partial x}\,dx\right)$$

in the opposite direction acting at the centre of BCC_1B_1.

Similarly the components of stress across the area ABB_1A_1 are (Y_x, Y_y, Y_z) and across DCC_1D_1 they are

$$\left(Y_x+\frac{\partial Y_x}{\partial y}\,dy,\ Y_y+\frac{\partial Y_y}{\partial y}\,dy,\ Y_z+\frac{\partial Y_z}{\partial y}\,dy\right).$$

For the area $ABCD$ the stress components are (Z_x, Z_y, Z_z) and for $A_1B_1C_1D_1$ they are

$$\left(Z_x+\frac{\partial Z_x}{\partial z}\,dz,\ Z_y+\frac{\partial Z_y}{\partial z}\,dz,\ Z+\frac{\partial Z_z}{\partial z}\,dz\right).$$

7.10. Conditions for equilibrium

Let ρ be the density of the medium and let $(\rho F_x, \rho F_y, \rho F_z)$ per unit volume be the components of the body forces acting on the element. It will be assumed that they act at the centre of the parallelepiped.

If the element is in statical equilibrium, then the forces acting on it must have (i) the sum of their components in any direction equal to zero and (ii) the sum of their moments about any line equal to zero. On summing parallel to the x-axis we have

$$X_x\,dydz-\left(X_x+\frac{\partial X_x}{\partial x}\,dx\right)dydz+Y_x\,dzdx-\left(Y_x+\frac{\partial Y_x}{\partial y}\,dy\right)dzdx+$$
$$+Z_x\,dxdy-\left(Z_x+\frac{\partial Z_x}{\partial z}\,dz\right)dxdy+\rho F_x\,dxdydz=0. \quad (1)$$

This evidently leads to

$$\left.\begin{aligned}
&\frac{\partial X_x}{\partial x}+\frac{\partial Y_x}{\partial y}+\frac{\partial Z_x}{\partial z}=\rho F_x.\\
\text{Similarly}\qquad &\frac{\partial X_y}{\partial x}+\frac{\partial Y_y}{\partial y}+\frac{\partial Z_y}{\partial z}=\rho F_z\\
\text{and}\qquad &\frac{\partial X_z}{\partial x}+\frac{\partial Y_z}{\partial y}+\frac{\partial Z_z}{\partial z}=\rho F_z.
\end{aligned}\right\} \quad (2)$$

On taking moments about a line parallel to the x-axis through the centre of the parallelepiped, it can be seen from Fig. VII. 2 that

$$Z_y\,dxdy\frac{dz}{2}+\left(Z_y+\frac{\partial Z_y}{\partial z}\,dz\right)dxdy\frac{dz}{2}-$$

$$-Y_z\,dzdx\frac{dy}{2}-\left(Y_z+\frac{\partial Y_z}{\partial y}\,dy\right)dzdx\frac{dy}{2}=0, \quad (3)$$

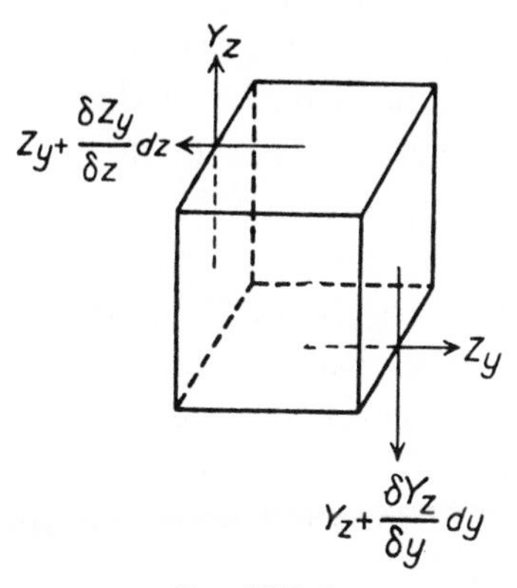

FIG. VII. 2.

leading to $Y_z = Z_y.$
Similarly we have $Z_x = X_z$
and $X_y = Y_x.$ (4)

7.11. Stress strain relations

The simple form of Hooke's law which applies to strained elastic rods has been generalized by Cauchy into a form which can be applied to strained elastic media in three dimensions. In this generalization it is assumed that each component of stress is linearly related to the six coefficients of strain. Denoting the components of strain in (7), § 7.8, by e_1, e_2, e_3, e_4, e_5, e_6, a typical Cauchy equation is

$$X_x = \sum_{i=1}^{6} a_i\,e_i, \quad (1)$$

where a_i $(i = 1, 2,..., 6)$ are constants depending upon the nature of the medium. There are six such equations involving altogether thirty-six constants, not all necessarily independent of each other. In the case of isotropic media†, i.e. media whose properties at a point are the same in all directions, it can be shown that all the constants can be expressed in terms of two independent parameters.‡

From the six equations such as (1) and the six equations given by the equilibrium conditions, (2) and (4), § 7.10, it is possible to eliminate the components of stress (or strain) and obtain six partial differential equations for the six components of strain (or stress). It can then be proved that there is only one solution of these equations which assumes given values, either of stress or strain, over a known boundary.§

The Cauchy assumptions (1) cannot be tested directly, but from the theory which rests upon them deductions can be made which are in close agreement with experiment.

In the case of the Saint-Venant torsion problem the relations between the components of stress and strain are practically self-evident and therefore we shall omit this part of the general theory. The part of the general theory which is of importance to us is the uniqueness theorem for given boundary conditions and this we shall assume without proof.

7.12. The Saint-Venant torsion problem

This problem, outlined in § 7.8, is treated here in more detail. A body, built of isotropic and elastic material, is in the shape of a prism bounded by a cylinder and two plane ends perpendicular to the generators of the cylinder. The body is twisted about an axis parallel to a generator by couples acting in the plane ends so that the relative rotation of two cross-sections is proportional to the distance between them, the rotation for unit distance being denoted by τ. If there are no external forces acting on the

† Media whose properties at a point are not the same in all directions are known as anisotropic. Examples of anisotropic media are crystals, drawn wire, and wood.

‡ A. E. H. Love, *Mathematical Theory of Elasticity*, § 66, p. 97; I. S. Sokolnikoff and R. D. Specht, *Mathematical Theory of Elasticity*, § 21, p. 59.

§ A. E. H. Love, loc. cit., § 118, p. 167; I. S. Sokolnikoff and R. D. Specht, loc. cit., § 27, p. 92; S. Timoshenko, *Theory of Elasticity*, § 64.

curved cylindrical part of the prism, then the problem of determining the internal stresses and strains admits of a unique solution which we now proceed to find.

For the case of a medium such as steel the problem can be simplified by using the following facts. For a stress of 6 tons weight per square inch in a steel rod the extension of length is 0·00046 inches per inch length of rod. Consequently the angle τ is very small in practice and the change in shape of the prism due to the strain is negligible. Again the mass of a cubic inch of steel is about $\frac{1}{4}$ lb., so that the body forces, which are due to gravity, are negligible in comparison with the stresses which occur in practice. We shall therefore write $\rho F_x = \rho F_y = \rho F_z = 0$ in (2), § 7.10.

When the prism is a portion of a right circular cylinder an exact solution can easily be found. Taking the axis of symmetry to be the z-axis (this being also the axis of twist), the point whose coordinates are (x, y, z) is displaced so that its coordinates are $(x+u, y+v, z+w)$, where

$$u = -\tau yz; \quad v = \tau zx; \quad w = 0. \tag{1}$$

Guided by this, Saint-Venant assumed for the case of the general cylindrical surface a displacement for which the functions u, v, w are defined as follows:

$$u = -\tau yz; \quad v = \tau zx; \quad w = \tau\phi(x, y), \tag{2}$$

where τ is the angle of twist per unit length of prism and $\phi(x, y)$ (for brevity denoted by ϕ) is independent of z. For such displacements points coplanar in the unstrained state are no longer coplanar when the strain is imposed.

In § 7.8 two neighbouring points in the unstrained medium, P and P_1, are displaced by the strain to Q and Q_1 respectively. The differences between the projections of QQ_1 and PP_1 on the coordinate axes are then evaluated and separated into two parts, one part due to rotation and the other to strain, (5), § 7.8. The part due to strain is given by the expressions (7), § 7.8, which, where the displacements are those of (2) above, reduce to the expressions (9), § 7.8, namely

$$\left.\begin{aligned} &\tfrac{1}{2}\tau\left(\frac{\partial\phi}{\partial x}-y\right)\zeta \\ &\tfrac{1}{2}\tau\left(\frac{\partial\phi}{\partial y}+x\right)\zeta \\ \tfrac{1}{2}\tau\left(\frac{\partial\phi}{\partial x}-y\right)\xi&+\tfrac{1}{2}\tau\left(\frac{\partial\phi}{\partial y}+x\right)\eta. \end{aligned}\right\} \tag{3}$$

The components of strain are the coefficients of ξ, η, ζ in (3).

We must now obtain the relations between the components of stress and the components of strain in (3). Note first that these

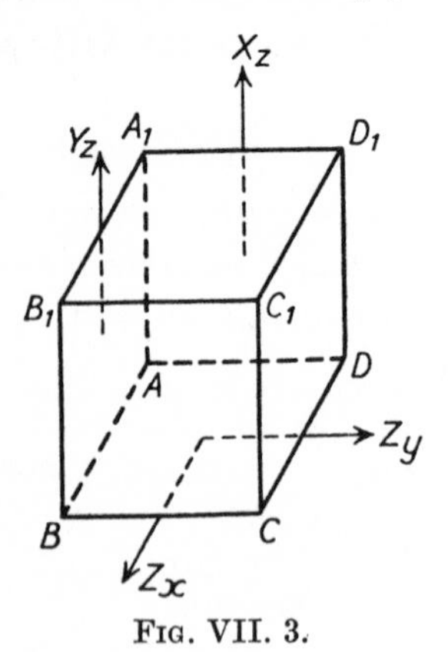

Fig. VII. 3.

relations are linear, that the medium is isotropic and that among the components of stress we must have $Y_z = Z_y$, $Z_x = X_z$, and $X_y = Y_x$, from (4), § 7.10. It is then evident from Fig. VII. 3 that all the conditions are satisfied by choosing

$$\left.\begin{aligned} X_x = Y_y = Z_z = X_y = Y_x = 0 \\ Z_x = X_z = \mu\tau\left(\frac{\partial\phi}{\partial x}-y\right) \\ Z_y = Y_z = \mu\tau\left(\frac{\partial\phi}{\partial y}+x\right) \end{aligned}\right\}, \tag{4}$$

where μ, a constant depending upon the nature of the medium, is known as the modulus of rigidity or the shear modulus.

On using the equilibrium equations (2), § 7.10, and writing $\rho F_x = \rho F_y = \rho F_z = 0$, as agreed at the commencement of this section, we see that the function ϕ must satisfy the Laplacian equation

$$\frac{\partial^2\phi}{\partial x^2}+\frac{\partial^2\phi}{\partial y^2} = 0. \tag{5}$$

The boundary condition which ϕ must satisfy is easily obtained. Over the cylindrical part of the prism there is no normal

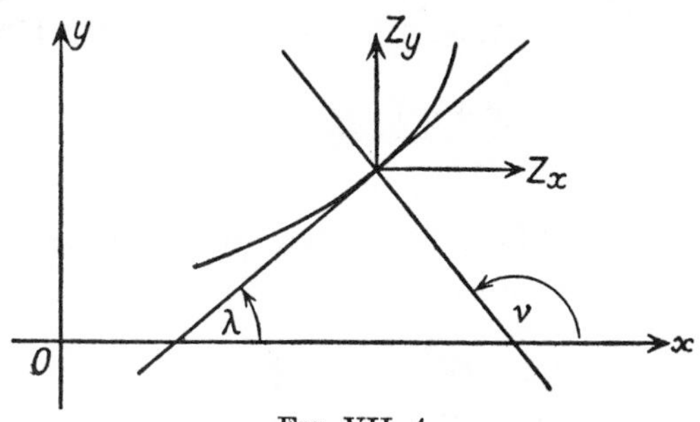

FIG. VII. 4.

stress, so that if ν is the angle between the x-axis and the normal (Fig. VII. 4) we must have

$$Z_x \cos\nu + Z_y \sin\nu = 0. \tag{6}$$

From (4) this reduces to

$$\frac{\partial\phi}{\partial x}\cos\nu+\frac{\partial\phi}{\partial y}\sin\nu = y\cos\nu - x\sin\nu. \tag{7}$$

Thus the Saint-Venant problem of torsion in an isotropic prism is reduced to the problem of finding a solution of (5) which satisfies (7) for all values of x and y on the boundary of a cross-section.

7.13. The variational form of Saint-Venant's problem

From theorem 10, § 3.8, and the results stated in § 3.9 it follows that the function $\phi(x, y)$, which satisfies (5), § 7.12, also minimizes the integral I, where

$$I = \iint \left\{\left(\frac{\partial\phi}{\partial x}\right)^2+\left(\frac{\partial\phi}{\partial y}\right)^2\right\} dxdy. \tag{1}$$

The area of integration for I is a cross-section of the prism.

In addition to making I a minimum $\phi(x, y)$ must also satisfy the boundary condition (7), § 7.12. This requirement can be expressed in a much more convenient form, as we now show.

If $f(z)$ is an analytic function of the complex variable†

$$z = x+iy$$

then, in some domain D, $df(z)/dz$ is independent of the value of $\arg dz$ as dz tends to zero. If $f(z) = \phi+i\psi$, where ϕ and ψ are real functions of x and y, then differentiation parallel to either coordinate axis must give the same result. Hence

$$\frac{\partial(\phi+i\psi)}{\partial x} = \frac{\partial(\phi+i\psi)}{i\,\partial y}. \tag{2}$$

On equating real and imaginary parts we obtain the well-known Riemann–Cauchy equations

$$\left.\begin{aligned}\frac{\partial\phi}{\partial x} &= \frac{\partial\psi}{\partial y}\\ \frac{\partial\phi}{\partial y} &= -\frac{\partial\psi}{\partial x}\end{aligned}\right\}. \tag{3}$$

From these we deduce that

$$\left.\begin{aligned}\frac{\partial^2\phi}{\partial x^2}+\frac{\partial^2\phi}{\partial y^2} &= 0\\ \text{and}\qquad \frac{\partial^2\psi}{\partial x^2}+\frac{\partial^2\psi}{\partial y^2} &= 0\end{aligned}\right\}. \tag{4}$$

Thus to every function $\phi(x,y)$ which satisfies (5), § 7.12, there corresponds a conjugate function $\psi(x,y)$ which also satisfies Laplace's equation and which is related to $\phi(x,y)$ by (3). The function $\psi(x,y)$ enables us to express the boundary condition (7), § 7.12, in a more convenient form.

Let λ denote the angle between the x-axis and the tangent to the boundary of a cross-section and, as in § 7.12, let ν denote the angle between the x-axis and the normal. Then (Fig. VII. 4)

$$\nu = \lambda+\tfrac{1}{2}\pi, \qquad \cos\nu = -\sin\lambda = -\frac{dy}{ds}, \qquad \sin\nu = \cos\lambda = \frac{dx}{ds}.$$

† Analytic functions of complex variables are treated in all works on mathematical analysis, e.g. E. T. Whittaker and G. N. Watson, *Modern Analysis*.

Combining these with (3) it is clear that the boundary condition (7), § 7.12, can be written in the form

$$-\frac{\partial\psi}{\partial y}\frac{dy}{ds}-\frac{\partial\psi}{\partial x}\frac{dx}{ds} = -y\frac{dy}{ds}-x\frac{dx}{ds}. \tag{5}$$

By integration we then have

$$\psi = \tfrac{1}{2}(x^2+y^2)+\text{constant} \tag{6}$$

at all points of the boundary. The value of the constant is not of great importance, since in practice only the derivatives of ψ are required. There is therefore no loss of generality in taking its value to be zero, as we shall do in future.

It is also evident that the boundary condition can be simplified still further by introducing a new function Ψ defined by the equation

$$\Psi = \psi-\tfrac{1}{2}(x^2+y^2), \tag{7}$$

so that on the boundary $\Psi = 0$. Throughout the interior the second equation of (4) shows that

$$\frac{\partial^2\Psi}{\partial x^2}+\frac{\partial^2\Psi}{\partial y^2}+2 = 0. \tag{8}$$

A further appeal to theorem 10, § 3.8, and the results of § 3.9 shows that Ψ minimizes the integral J, where

$$J = \iint \left\{\left(\frac{\partial\Psi}{\partial x}\right)^2+\left(\frac{\partial\Psi}{\partial y}\right)^2-4\Psi\right\} dxdy. \tag{9}$$

As before, the area of integration is a cross-section of the prism.

The Saint-Venant torsion problem is therefore equivalent to that of finding a function Ψ which minimizes J and which vanishes on the boundary of the prism. Having determined Ψ we obtain ψ from (7) and ϕ from (3). The stresses and strains in the medium can then be obtained from (4) and (2), § 7.12.

Example 1. Consider the case of a prism with circular cross-section of radius a, the centre of the circle being on the axis of twist. Take $\Psi = \alpha(a^2-x^2-y^2)$ and show that J is a minimum when $\alpha = \frac{1}{2}$. Deduce the values of ϕ, Z_x, and Z_y.

Example 2. Deduce the minimum property of J, equation (9), from that of I, equation (1).

EXAMPLE 3. For the prism with the elliptical boundary whose equation is $x^2/a^2+y^2/b^2 = 1$ take $\Psi = \alpha\left(\frac{x^2}{a^2}+\frac{y^2}{b^2}-1\right)$ and find the value of α which makes J a minimum. Deduce that†

$$\phi = -\frac{a^2-b^2}{a^2+b^2}xy.$$

7.14. The torsion of beams with rectangular cross-section

The Saint-Venant problem for the beam with cross-section bounded by the rectangle $x = \pm a$, $y = \pm b$ has been solved in terms of Fourier series.‡ In applying the Rayleigh–Ritz method to this problem the simplest approximating function to Ψ is

$$\Psi_1 = \alpha(a^2-x^2)(b^2-y^2), \tag{1}$$

where α is to be determined. Ψ_1 evidently vanishes on the boundary. On substituting in the integrand of (9), § 7.13, and integrating between the limits $-a$ and a for x, $-b$ and b for y we obtain

$$J = \frac{64}{45}a^3b^3\{2\alpha^2(a^2+b^2)-5\alpha\}. \tag{2}$$

Consequently J is a minimum when

$$\alpha = \frac{5}{4(a^2+b^2)}. \tag{3}$$

For comparison with the exact answer we make use of the quantity known as the torsional rigidity. If R is the torsional rigidity and τ is the twist per unit length of prism, then $R\tau$ is equal to the moment, about the axis of twist, of the forces acting over a cross-section of the prism. From Fig. VII. 3 we have

$$\begin{aligned} R\tau &= \iint (Z_y x - Z_x y)\,dxdy \\ &= \mu\tau \iint \left\{\left(\frac{\partial\phi}{\partial y}+x\right)x-\left(\frac{\partial\phi}{\partial x}-y\right)y\right\}\,dxdy, \end{aligned} \tag{4}$$

on using (4), § 7.12. From (3) and (7), § 7.13, we deduce that

$$R = -\mu \iint \left(\frac{\partial\Psi}{\partial y}y+\frac{\partial\Psi}{\partial x}x\right)\,dxdy. \tag{5}$$

† This result is the correct answer, see A. E. H. Love, loc. cit., p. 305.
‡ A. E. H. Love, loc. cit., § 221, p. 305; S. Timoshenko, loc. cit., p. 245.

A simple transformation enables us to put this into a much more convenient form for practical use. Evidently we have

$$R = \mu \iint \left\{2\Psi - \frac{\partial(x\Psi)}{\partial x} - \frac{\partial(y\Psi)}{\partial y}\right\} dxdy. \tag{6}$$

But from the lemma of § 3.7 we have

$$\iint \left\{\frac{\partial(x\Psi)}{\partial x} + \frac{\partial(y\Psi)}{\partial y}\right\} dxdy = \int (x\Psi\, dy - y\Psi\, dx), \tag{7}$$

where the left-hand integral is taken over the area of a cross-section and the right-hand one over the boundary curve. But on the boundary we have $\Psi = 0$, hence each side of (7) must vanish and (6) reduces to

$$R = 2\mu \iint \Psi\, dxdy. \tag{8}$$

Taking the simple case of the square, where $2a = 2b = l$, the length of a side, we have $\alpha = \frac{5}{8}a^2$ from (3), so that the approximate form for Ψ_1 becomes $5(a^2-x^2)(a^2-y^2)/8a^2$. On inserting this in (8) we get $R = \mu 20a^4/9 = 0{\cdot}1389\mu l^4$. The exact result† is $0{\cdot}1406\mu l^4$, showing how close the approximation is although the form assumed for Ψ_1 is a function of the simplest kind.

EXAMPLE 4. Calculate the value of R for the rectangular beam whose sides are of length $2a$, $2b$. Show that when $b = 10a$ the approximate value of R is $2{\cdot}75\mu l^4$, where $l = 2a$. (The exact value is $3{\cdot}12\mu l^4$.)

EXAMPLE 5. For a beam with square cross-section take $\Psi_1 = \alpha\cos(\pi x/l)\cos(\pi y/l)$, where l is equal to the length of a side, as the function approximating to Ψ. Show that J, (9), § 7.13, is minimized when $\alpha = 16l^2/\pi^4$ and that the torsional rigidity is approximately $0{\cdot}1331\mu l^4$.

There is a simple relation between R, the torsional rigidity, and the minimum value of J, the integral of (9), § 7.13. Writing J_m for the minimum value of J the relation is

$$R = -\mu J_m. \tag{9}$$

† I. S. Sokolnikoff and R. D. Specht, loc. cit., p. 309; S. Timoshenko, loc. cit., p. 249, $R = M_t/\theta$.

To prove this, from the definition of J in § 7.13 it follows that

$$J = \iint \left\{\frac{\partial}{\partial x}\left(\Psi\frac{\partial\Psi}{\partial x}\right)+\frac{\partial}{\partial y}\left(\Psi\frac{\partial\Psi}{\partial y}\right)\right\} dxdy-$$
$$-\iint \left\{\frac{\partial^2\Psi}{\partial x^2}+\frac{\partial^2\Psi}{\partial y^2}+4\right\}\Psi\, dxdy. \quad (10)$$

From the lemma of § 3.7 the first integral is equal to

$$\int \Psi\left\{\frac{\partial\Psi}{\partial x}\,dy-\frac{\partial\Psi}{\partial y}\,dx\right\}$$

taken round the boundary where $\Psi = 0$. Hence the first integral of (10) vanishes. For the second integral we note that J is a minimum if Ψ is chosen to satisfy equation (8), § 7.13, so that (10) then reduces to

$$J_m = -2\iint \Psi\, dxdy. \quad (11)$$

The proof of (9) is completed by comparing (8) and (11).

7.15. Upper bounds for the integral J, (9), § 7.13

In applying the Rayleigh–Ritz method to the problem of minimizing the integral J, the unknown function Ψ is replaced by a series of known functions multiplied by unknown parameters. Let such an approximating series be denoted by Ψ_n and the corresponding value of the integral by J_n. In this section we shall attempt to estimate the degree of approximation attained by these methods. We prove that if Ψ_n vanishes on the boundary then

$$J_n \geqslant J. \quad (1)$$

Writing

$$D = \Psi-\Psi_n \quad (2)$$

and using the definition of J given by equation (9), § 7.13, we have

$$J_n-J = \iint \left\{\left(\frac{\partial\Psi_n}{\partial x}\right)^2+\left(\frac{\partial\Psi_n}{\partial y}\right)^2-4\Psi_n\right\} dxdy-$$
$$-\iint \left\{\left(\frac{\partial\Psi}{\partial x}\right)^2+\left(\frac{\partial\Psi}{\partial y}\right)^2-4\Psi\right\} dxdy \quad (3)$$
$$= -2\iint \left\{\frac{\partial\Psi}{\partial x}\frac{\partial D}{\partial x}+\frac{\partial\Psi}{\partial y}\frac{\partial D}{\partial y}\right\} dxdy+4\iint D\, dxdy+$$
$$+\iint \left\{\left(\frac{\partial D}{\partial x}\right)^2+\left(\frac{\partial D}{\partial y}\right)^2\right\} dxdy, \quad (4)$$

where, as always, the integration is taken over the cross-section of the prism.

Since Ψ and Ψ_n both vanish on the boundary the same must be true of D. Hence on integrating by parts with respect to x it follows that

$$\iint \frac{\partial \Psi}{\partial x}\frac{\partial D}{\partial x}\,dxdy = -\iint D\frac{\partial^2 \Psi}{\partial x^2}\,dxdy. \tag{5}$$

Similarly, on integrating by parts with respect to y, we have

$$\iint \frac{\partial \Psi}{\partial y}\frac{\partial D}{\partial y}\,dxdy = -\iint D\frac{\partial^2 \Psi}{\partial y^2}\,dxdy. \tag{6}$$

But we also know that Ψ satisfies equation (8), § 7.13, namely

$$\frac{\partial^2 \Psi}{\partial x^2}+\frac{\partial^2 \Psi}{\partial y^2}+2 = 0. \tag{7}$$

Consequently the sum of the first two integrals on the right-hand side of (4) must be zero.

But the integrand of the third integral of (4) is obviously positive, so that we have

$$J_n \geqslant J \tag{8}$$

as required.

This result shows that the Rayleigh–Ritz method furnishes a series of upper bounds for the value of J. At the same time, from (9), § 7.14, we obtain a series of lower bounds for the torsional rigidity R.

7.16. Lower bounds for the integral J, (9), § 7.13

A method of finding lower bounds for the integral J, and upper bounds for the torsional rigidity R, was evolved by Trefftz.†

In this method we revert to the function ψ introduced in § 7.13. The properties of ψ are as follows:

(i) Throughout a cross-section

$$\frac{\partial^2 \psi}{\partial x^2}+\frac{\partial^2 \psi}{\partial y^2} = 0. \tag{1}$$

† E. Trefftz, 'Konvergenz und Fehlerabschätzung beim Ritz'schen Verfahren', *Mathematische Annalen*, **100** (1928), 503–21; 'Ein Gegenstück zum Ritz'schen Verfahren', *Proc. Second International Congress for Applied Mechanics* 1927, pp. 131–7.

(ii) Over the boundary of a cross-section

$$\psi = \tfrac{1}{2}(x^2+y^2). \tag{2}$$

(iii) The relation between the function Ψ of the previous section and ψ is

$$\Psi = \psi-\tfrac{1}{2}(x^2+y^2). \tag{3}$$

In §§ 7.14 and 7.15 we approximated to Ψ by means of a function Ψ_n which satisfied the same boundary conditions as Ψ, i.e. Ψ_n vanished on the boundary. In the Trefftz method we approximate to the function ψ by means of a function ψ_n which satisfies Laplace's equation (1). Of these two methods the first leads to a series of upper bounds for J, as we have already shown, and the second leads to a series of lower bounds for J, as we now proceed to prove.

We must first express the integral J, (9), § 7.13, in terms of the function ψ and for this purpose we introduce the integral K defined by

$$K = \iint \left\{\left(\frac{\partial\psi}{\partial x}\right)^2+\left(\frac{\partial\psi}{\partial y}\right)^2\right\} dxdy. \tag{4}$$

Here, as in all subsequent double integrals of this section, the area of integration is over a cross-section of the prism.

The relation we require is

$$J = K-\iint (x^2+y^2)\, dxdy. \tag{5}$$

To prove this result by (3) the right-hand side of (5) is equal to

$$\iint \left\{\left(\frac{\partial\Psi}{\partial x}+x\right)^2+\left(\frac{\partial\Psi}{\partial y}+y\right)^2\right\} dxdy-\iint (x^2+y^2)\, dxdy$$

$$= J+\iint \left\{4\Psi+2x\frac{\partial\Psi}{\partial x}+2y\frac{\partial\Psi}{\partial y}\right\} dxdy \tag{6}$$

$$= J+2\iint \left\{\frac{\partial(x\Psi)}{\partial x}+\frac{\partial(y\Psi)}{\partial y}\right\} dxdy. \tag{7}$$

On using the lemma of § 3.7 we have

$$2\iint \left\{\frac{\partial(x\Psi)}{\partial x}+\frac{\partial(y\Psi)}{\partial y}\right\} dxdy = 2\int \Psi(x\, dy-y\, dx), \tag{8}$$

where the right-hand integral is taken round the boundary. But on the boundary $\Psi = 0$ (equations (2) and (3)), so that the integral in (7) vanishes. Consequently the right-hand side of (5) is identically equal to J.

From (5) the difference between J and K is equal to the moment of inertia of a cross-section about the axis of twist. Since this is constant we see that from a set of lower bounds for K it is a simple matter to deduce a set of lower bounds for J.

We now proceed to find ψ by means of an approximating function ψ_n defined by the series

$$\psi_n = \sum_m^n a_m f_m(x,y), \tag{9}$$

where each function $f_m(x,y)$ satisfies Laplace's equation in two dimensions, i.e.

$$\frac{\partial^2 f_m(x,y)}{\partial x^2} + \frac{\partial^2 f_m(x,y)}{\partial y^2} = 0 \tag{10}$$

($m = 1, 2,..., n$). From the results of § 7.13 the functions f_m can be chosen from the real or imaginary parts of any convenient function of the complex variable z ($= x+iy$, where $i = \surd(-1)$), and in particular from the polynomial real or imaginary parts of $(x+iy)^n$. For with such choice of f_m the function ψ_n must also satisfy Laplace's equation.

Let

$$d = \psi - \psi_n; \tag{11}$$

then we shall write

$$K_n = \iint \left\{\left(\frac{\partial \psi_n}{\partial x}\right)^2 + \left(\frac{\partial \psi_n}{\partial y}\right)^2\right\} dxdy \tag{12}$$

and

$$K_d = \iint \left\{\left(\frac{\partial d}{\partial x}\right)^2 + \left(\frac{\partial d}{\partial y}\right)^2\right\} dxdy. \tag{13}$$

The relation between K, K_n, and K_d reveals the Trefftz method most clearly. We have from (4)

$$K = \iint \left\{\left(\frac{\partial(\psi_n+d)}{\partial x}\right)^2 + \left(\frac{\partial(\psi_n+d)}{\partial y}\right)^2\right\} dxdy \tag{14}$$

$$= K_n + K_d + 2\iint \left\{\left(\frac{\partial d}{\partial x}\right)\left(\frac{\partial \psi_n}{\partial x}\right) + \left(\frac{\partial d}{\partial y}\right)\left(\frac{\partial \psi_n}{\partial y}\right)\right\} dxdy. \tag{15}$$

So far the coefficients a_m ($m = 1, 2,..., n$) in (9) are arbitrary. The Trefftz method depends upon the fact that if these coefficients are chosen to make K_d stationary, then the integral in (15) must vanish, leaving us with the equation

$$K = K_n + K_d. \tag{16}$$

To prove this, the values of a_m for which K_d is stationary are obtained by solving the n equations

$$\frac{\partial K_d}{\partial a_m} = 0 \quad (m = 1, 2, \ldots, n). \tag{17}$$

But from (9) and (11) we have

$$\frac{\partial d}{\partial a_m} = -f_m(x, y) = -f_m. \tag{18}$$

Hence from (13) and (17) we have

$$\iint \left\{ \frac{\partial d}{\partial x} \frac{\partial f_m}{\partial x} + \frac{\partial d}{\partial y} \frac{\partial f_m}{\partial y} \right\} dxdy = 0 \tag{19}$$

$(m = 1, 2, \ldots, n)$. These n equations suffice to determine the n parameters a_m.

Multiply (19) by a_m, sum for all values of m from 1 to n and use (9). It follows that the integral in (15) vanishes when K_d is stationary, thus justifying the statements which led up to (16).

From (13) $K_d \geqslant 0$, so that when K_d is stationary we see from (16) that

$$K_n \leqslant K. \tag{20}$$

In general K_d has only one stationary value and as it is positive semi-definite this must be a minimum. The value of K_n will then be as near to K as the choice of the functions $f_m(x, y)$ permits.

7.17. Applications of the Trefftz method

Although equations (19), § 7.16, enable us to evaluate the parameters a_m, yet they are not in the form most convenient for practical application. By means of the lemma of § 3.7 the integrals in these equations can be transformed into boundary integrals which are much easier to evaluate.

From (2) and (11), § 7.16, we have on the boundary

$$d = \tfrac{1}{2}(x^2 + y^2) - \psi_n. \tag{1}$$

But from (10), § 7.16, we can write (19), § 7.16, in the form

$$\iint \left\{ \frac{\partial}{\partial x}\left(d \frac{\partial f_m}{\partial x}\right) + \frac{\partial}{\partial y}\left(d \frac{\partial f_m}{\partial y}\right) \right\} dxdy = 0. \tag{2}$$

Combining this with the lemma of § 3.7 it follows that (19), § 7.16, can be replaced by

$$\int \{\tfrac{1}{2}(x^2+y^2)-\psi_n\}\left\{\frac{\partial f_m}{\partial x}\,dy-\frac{\partial f_m}{\partial y}\,dx\right\}=0 \tag{3}$$

($m = 1, 2,..., n$), where the integration is taken round the boundary. These equations are not identities since ψ_n satisfies Laplace's equation (see (9) and (10), § 7.16), and does not assume the value $\frac{1}{2}(x^2+y^2)$ on the boundary.

If the functions in the integrand of (3) have the property of completeness (see § 7.5), then as n tends to infinity it may reasonably be expected that ψ_n will tend to ψ, but an analytical discussion of this point presents great difficulty. Sufficient conditions to ensure the validity of $\lim_{n\to\infty}\psi_n = \psi$ are of great importance, but in practice a knowledge of the magnitude of the error $\psi-\psi_n$ is often even more important. For this reason we shall omit a theoretical discussion and rely upon the upper and lower bounds established in §§ 7.15 and 7.16 whenever calculations of torsional rigidity or other quantities of interest are required.

There still remains one point for discussion. It is often convenient to take f_1 to be a constant and in fact many complete sets of functions commence with a constant. Since the integrals for K_d involve only derivatives of ψ and ψ_n, the value of f_1 has no influence upon the stationary values of K_d. Consequently equations (2) start from the value $m = 2$ and do not contain a_1. Clearly the best choice of a_1 is one which makes ψ_n attain as near as possible the value $\frac{1}{2}(x^2+y^2)$ on the boundary, and this is done by making the mean value of the difference vanish. We then have the equation

$$\int \{\tfrac{1}{2}(x^2+y^2)-\psi_n\}\,ds = 0, \tag{4}$$

where the integration is taken round the boundary of a cross-section. Equations (3) for $m = 2, 3,..., n$ together with (4) give us sufficient equations to determine all the parameters a_m.

To illustrate these results we consider once again the torsion of a prism of square cross-section. From the nature of the problem the functions $f_m(x,y)$ must be of even degree and symmetrical in x and y in addition to satisfying Laplace's equation. We therefore choose

$$f_1 = 1; \quad f_2 = \text{real part of } (x+iy)^4 = x^4-6x^2y^2+y^4, \tag{5}$$

and so assume that

$$\psi_2 = \alpha+\beta(x^4-6x^2y^2+y^4), \tag{6}$$

where α and β are to be determined from equations (3) and (4). From the symmetry of the problem it is sufficient to take the boundary integrals along one side of the square, say the side parallel to the y-axis where $x = a$. Equation (4) gives us

$$\int\limits_{-a}^{a} \{\tfrac{1}{2}(a^2+y^2)-\alpha-\beta(a^4-6a^2y^2+y^4)\}\, dy = 0, \tag{7}$$

which leads to $$15\alpha-12a^4\beta = 10a^2. \tag{8}$$

Equation (3), with $m = 2$, gives us

$$\int\limits_{-a}^{a} \{\tfrac{1}{2}(a^2+y^2)-\alpha-\beta(a^4-6a^2y^2+y^4)\}\frac{\partial}{\partial x}(x^4-6x^2y^2+y^4)\, dy = 0, \tag{9}$$

where $x = a$ after the differentiation has been performed. On using (7) it is evident that (9) reduces to

$$\int\limits_{-a}^{a} \{\tfrac{1}{2}(a^2+y^2)-\alpha-\beta(a^4-6a^2y^2+y^4)\}y^2\, dy = 0. \tag{10}$$

This gives us $$35\alpha-76a^4\beta = 28a^2. \tag{11}$$

From (8) and (11) we have

$$\alpha = \frac{53}{90}a^2; \qquad \beta = -\frac{7}{72a^2}, \tag{12}$$

and so $$\psi_2 = \frac{53}{90}a^2-\frac{7}{72a^2}(x^4-6x^2y^2+y^4). \tag{13}$$

From (3), § 7.16, the corresponding approximation to Ψ, namely Ψ_2, is given by

$$\Psi_2 = \frac{53}{90}a^2-\frac{7}{72a^2}(x^4-6x^2y^2+y^4)-\tfrac{1}{2}(x^2+y^2). \tag{14}$$

We now apply these results to obtain an approximation to the value of R, the torsional rigidity defined in § 7.14. For this purpose we make use of equation (8), § 7.14, namely

$$R = 2\mu \iint \Psi \, dxdy. \tag{15}$$

Substituting Ψ_2 for Ψ and evaluating we obtain the desired approximation to R. If $2a = l$ the result is

$$R = \mu \frac{19}{135} l^4 = 0{\cdot}1407\mu l^4. \tag{16}$$

The Trefftz method furnishes us with the following information: (i) from (20), § 7.16, lower bounds for K; (ii) from (5), § 7.16, lower bounds for J; and (iii) from (9), § 7.14, upper bounds for R. The Rayleigh–Ritz method gives us: (i) from (8), § 7.15, upper bounds for J, and (ii) from (9), § 7.14, lower bounds for R. On comparing (16) above with the approximation for R obtained by the Rayleigh–Ritz method in § 7.14 we see that without finding the exact value (namely $0{\cdot}1406\mu l^4$) we have proved that R must lie between $0{\cdot}1389\mu l^4$ and $0{\cdot}1407\mu l^4$. Evidently for most practical purposes such close limits will suffice.

7.18. Galerkin's method

An ingenious alternative to the Rayleigh–Ritz method has been evolved by Galerkin.† To explain the ideas involved we must refer back to some of the work of Chapter III. In §§ 3.6, 3.7, and 3.8 we investigated conditions which ensured that

$$I = \iint F(x, y, z, p, q) \, dxdy \tag{1}$$

should have a stationary value. In this integral F is a known functional form, x and y are the independent variables, z is the dependent variable, $p = \partial z/\partial x$ and $q = \partial z/\partial y$.

Let $z = s(x, y)$ be the equation of the surface for which (1) is stationary and $z = s(x, y) + \epsilon\eta(x, y)$ be the equation of a neighbouring surface. Let the values of (1) for these two surfaces

† The original paper is in Russian but the ideas have been expounded by Professor W. J. Duncan in a series of papers in the *Philosophical Magazine*. The paper by B. G. Galerkin is 'Series Solutions of Some Problems of Elastic Equilibrium of Rods and Plates', *Vestnik Inghenerov*, **1** (1915), 879–908. See also W. J. Duncan 'Application of Galerkin's Method to the Torsion and Flexure of Cylinders and Prisms', *Phil. Mag.*, Ser. 7, **25** (1938), 636–49.

be I and $I+\delta I$ respectively. It was established in §§ 3.6 and 3.8 that for sufficiently small ϵ

$$\delta I = \epsilon \iint \eta(x,y)\left\{\frac{\partial F}{\partial s}-\frac{\partial}{\partial x}\left(\frac{\partial F}{\partial p}\right)-\frac{\partial}{\partial y}\left(\frac{\partial F}{\partial q}\right)\right\} dxdy+O(\epsilon^2). \quad (2)$$

For a stationary value of I the coefficient of ϵ must vanish and the arbitrary nature of $\eta(x,y)$ enables us to deduce further, as in § 1.4, that z must satisfy the equation

$$\frac{\partial F}{\partial z}-\frac{\partial}{\partial x}\left(\frac{\partial F}{\partial p}\right)-\frac{\partial}{\partial y}\left(\frac{\partial F}{\partial q}\right) = 0. \quad (3)$$

The Galerkin method of minimizing (1) consists in finding an approximation to the solution of (3). If z_n is such an approximation it is assumed that

$$z_n = \sum_{m=1}^{n} a_m f_m(x,y), \quad (4)$$

where the functions $f_m(x,y)$ are known and are chosen according to the nature of the problem and the quantities a_m $(m = 1, 2,..., n)$ are a group of n parameters whose values remain to be determined. Galerkin determines these values by a method which depends upon the fact that the integral on the right-hand side of (2) must vanish for any arbitrary function $\eta(x,y)$. He chooses $\eta(x,y)$ so that

$$\eta(x,y) = f_m(x,y), \quad (5)$$

where m can have any value from 1 to n, and writes

$$\iint f_m(x,y)\left\{\frac{\partial F}{\partial z_n}-\frac{\partial}{\partial x}\left(\frac{\partial F}{\partial p}\right)-\frac{\partial}{\partial y}\left(\frac{\partial F}{\partial q}\right)\right\} dxdy = 0 \quad (6)$$

$(m = 1, 2,..., n)$. These n equations suffice to determine the n parameters a_m.

Since z_n is not an exact solution of the Eulerian equation (3), the coefficient of ϵ in (2) does not, in general, vanish for any arbitrary variation $\eta(x,y)$. It vanishes only when $\eta(x,y)$ is chosen in accordance with (5). If the infinite sequence $f_m(x,y)$, $(m = 1, 2,...)$, possesses the property of completeness, then an arbitrary function $\eta(x,y)$, subject to some general restriction such as piecewise continuity, can be expressed in the form

$$\eta(x,y) = \sum_{m=1}^{\infty} b_m f_m(x,y). \quad (7)$$

Under such circumstances it may reasonably be expected that as n tends to infinity z_n will tend to a solution of (3), but the analysis required to establish such a result involves considerations of great complexity into which we do not enter here.

If we write
$$e_n = \frac{\partial F}{\partial z_n} - \frac{\partial}{\partial x}\left(\frac{\partial F}{\partial p}\right) - \frac{\partial}{\partial y}\left(\frac{\partial F}{\partial q}\right) \tag{8}$$
and if z_n does tend to a solution of (3) as n tends to infinity, then e_n must simultaneously tend to zero. Thus e_n may be used to measure the degree of approximation attained.

Like all methods of approximation success depends largely upon the choice of the functions $f_m(x,y)$. As far as possible they should be chosen to satisfy the same conditions as are prescribed for z. For example, if z is to satisfy given boundary conditions it is advisable to choose each of the functions $f_m(x,y)$ so as to satisfy these conditions also.

For the torsion of a prism the discussion of § 7.13 shows that we must minimize the integral
$$J = \iint \left\{\left(\frac{\partial \Psi}{\partial x}\right)^2 + \left(\frac{\partial \Psi}{\partial y}\right)^2 - 4\Psi\right\} dxdy, \tag{9}$$
where the function Ψ vanishes on the boundary of the area of integration. To apply the Galerkin method we write $z = \Psi$, $p = \partial\Psi/\partial x$, $q = \partial\Psi/\partial y$ in (6), and assume that Ψ_n, an approximation to Ψ, is given by
$$\Psi_n = \sum_{m=1}^{n} a_m f_m(x,y). \tag{10}$$
The Galerkin equations for a_m are then
$$\iint f_m(x,y)\left\{\frac{\partial^2 \Psi_n}{\partial x^2} + \frac{\partial^2 \Psi_n}{\partial y^2} + 2\right\} dxdy = 0, \tag{11}$$
where, as usual, the area of integration is over a cross-section of the prism. The functions $f_m(x,y)$ are chosen so as to vanish on the boundary if possible.

For the case of the beam whose cross-section is bounded by the rectangle $x = \pm a$, $y = \pm b$ we assume, as in § 7.14, that
$$\Psi_1 = \alpha(a^2 - x^2)(b^2 - y^2). \tag{12}$$

Equation (11) then gives us

$$\int_{-b}^{b}\int_{-a}^{a} (a^2-x^2)(b^2-y^2)\{2\alpha(x^2+y^2-a^2-b^2)+2\}\,dxdy = 0. \quad (13)$$

On evaluation we obtain $\alpha = \dfrac{5}{4(a^2+b^2)}$ as in the Rayleigh–Ritz application in § 7.14. The Galerkin method is usually quicker in practice than the Rayleigh–Ritz method as this example shows.

EXAMPLE 6. Apply the Galerkin method to the torsion of a beam of square cross-section using the assumption that

$$\Psi_2 = \alpha(a^2-x^2)(b^2-y^2)+\beta(a^4-x^4)(b^4-y^4).$$

7.19. Variations of the Rayleigh–Ritz and Galerkin methods

Numerous variations of the Rayleigh–Ritz and Galerkin methods have been proposed in recent times. In most cases the function sought is approximated to by means of an expansion of the form $\sum_{m=1}^{n} a_m f_m(x,y)$, where the functions $f_m(x,y)$ are chosen according to the nature of the problem and the quantities a_m remain to be determined.

In one variation the quantities a_m are taken to be functions of x so that the n simultaneous algebraical equations which occur in the Rayleigh–Ritz, Trefftz, and Galerkin methods are replaced by n simultaneous differential equations. From these a_m ($m = 1, 2,..., n$) can be determined as functions of x.†

In the Galerkin method the function $\eta(x,y)$ is chosen according to equation (5), § 7.18. In some of the variations of the method a different choice is taken for $\eta(x,y)$, for example, in one case $\eta(x,y)$ is chosen to have the value 1 over some parts of the cross-sectional area of the prism and the value zero over the remaining parts.‡

† L. V. Kantorovitch, *Bulletin of the Academy of Sciences of U.S.S.R.* No. 5 (1903); *Applied Mathematics and Mechanics*, **6** (1942), 31–40.

‡ C. B. Biezeno and J. J. Koch, *De Ingenieur*, **38** (1923), 25–36.

Another method consists in setting up the usual approximate form for z and then finding the values of a_m so that

$$\iint e_n^2 \, dxdy \tag{1}$$

is a minimum, where e_n is defined by (8), § 7.18. As usual the area of integration is a cross-section of the prism. As pointed out in § 7.18, e_n is a measure of the degree of approximation, so that minimizing (1) makes the approximation as close as possible. The equations for minimizing (1) are

$$\iint e_n \frac{\partial e_n}{\partial a_m} \, dxdy \tag{2}$$

$(m = 1, 2,..., n)$. We thus obtain n equations to determine the n parameters a_m.

The success of all these methods depends largely upon the choice of the first approximation Ψ_1 or z_1. If this is carefully chosen a high degree of accuracy can often be attained with comparatively little numerical calculation, but if the choice is not a fortunate one a considerable amount of labour may follow. The method of finite differences has been developed in order to overcome the undesirable consequences caused by an unsuitable choice of the first approximating function. In this method the area of integration is replaced by a set of points and the differential equation for minimizing the integral is replaced by one or more difference equations. Accounts of this and similar methods can be found in various works,† and a very interesting and comprehensive review of the subject together with a large list of references is given by R. Courant‡ in an address delivered to the American Mathematical Society.

† I. J. Sokolnikoff and R. D. Specht, loc. cit., § 73.

‡ R. Courant, 'Variational Methods for the Solution of Problems of Equilibrium and Vibration', *Bulletin of the American Mathematical Society*, **49** (1943), 1–23.

CHAPTER VIII

INTEGRALS WITH VARIABLE END POINTS HILBERT'S INTEGRAL

8.1. Introduction

THE analyses of the previous chapters apply only to integrals whose paths of integration have fixed end points. In this chapter we shall deal with cases where the end points can be displaced along prescribed curves.

We confine ourselves to the case of one independent and one dependent variable only, but we shall use somewhat more general arguments than those employed in Chapters I and II.

It will appear later that the vanishing of the first variation gives the same characteristic equation as found in § 1.4 and in addition gives equations known as transversality conditions. These conditions enable us to determine the constants of the solution although the coordinates of the end points are not known. The study of the second variation is somewhat complex. It leads to the introduction of focal points, which are analogous to the conjugate points of § 2.6 but have a wider field of application.

8.2. First variation with one end point variable

Let
$$I = \int_a^b F(x, y, y')\, dx \tag{1}$$

where $y' = dy/dx$ and A and B, the end points of the arc of integration, have abscissae a and b respectively (Fig. VIII. 1). In finding the stationary value of I we shall not only vary the arc AB but we shall also allow B to move along the curve Γ_2 whose equation is
$$y = g_2(x). \tag{2}$$

In this investigation the point A will be kept fixed, but the results obtained are easily generalized to allow for the case when A can be displaced along the curve $y = g_1(x)$.

In § 1.3, for given x, the ordinate was varied from $y = s(x)$, the stationary curve or extremal, to $y = s(x)+\epsilon t(x)$. In this chapter we shall vary the ordinate in a more general manner. For the extremal we take $y = y(x, 0)$ (arc AB in Fig. VIII. 1) and for the varied curve $y = y(x, \epsilon)$ (arc AB' in Fig. VIII. 1).

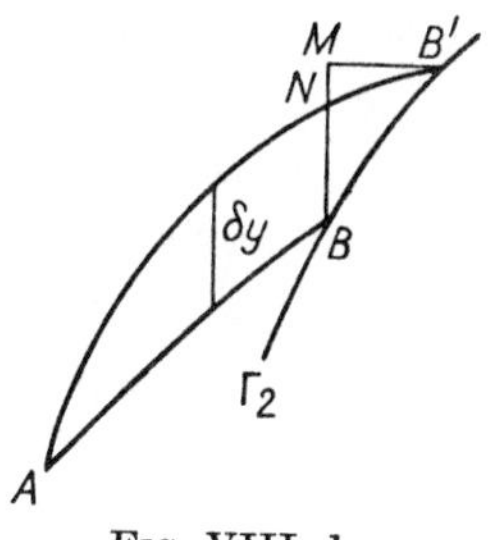

FIG. VIII. 1.

The stationary case occurs when $\epsilon = 0$. Now in the varied case I will be a function of ϵ and the usual methods of the calculus then give the following results: for a maximum at $\epsilon = 0$ we have

$$\frac{dI}{d\epsilon} = 0, \quad \frac{d^2I}{d\epsilon^2} < 0 \text{ and for a minimum } \frac{dI}{d\epsilon} = 0, \quad \frac{d^2I}{d\epsilon^2} > 0.$$

If B is displaced along Γ_2 to B', whose abscissa is $b+dx_b$, then

$$\delta I = \int_a^{b+dx_b} F(x, y+\delta y, y'+\delta y')\, dx - \int_a^b F(x, y, y')\, dx, \tag{3}$$

where $dx_b = MB'$ in Fig. VIII. 1 and $\delta y = y(x, \epsilon)-y(x, 0)$. The part of (3) which contains dx_b, δy, $\delta y'$ linearly will, as usual, be called the first variation and will be denoted by δI_1. We have

$$\delta I_1 = F_b\, dx_b + \int_a^b \left\{\frac{\partial F}{\partial y}\,\delta y + \frac{\partial F}{\partial y'}\,\delta y'\right\} dx, \tag{4}$$

where F is an abbreviation for $F(x, y, y')$ and F_b is the value of F at $x = b$.† If $\delta y' = d(\delta y)/dx$, then the second term in the

† (4) can be obtained by differentiating $\int_a^b F(x, y, y')\, dx$ with respect to ϵ, allowing for the fact that b is a function of ϵ.

integral of (4) can be integrated by parts, giving us

$$\delta I_1 = F_b\, dx_b + \left(\frac{\partial F}{\partial y'}\right)_b \delta y_b + \int_a^b \delta y \left\{\frac{\partial F}{\partial y} - \frac{d}{dx}\left(\frac{\partial F}{\partial y'}\right)\right\} dx. \tag{5}$$

Here $\delta y_b = y(b,\epsilon) - y(b,0)$ and the subscript b denotes that the variables have values corresponding to $x = b$. Since A is a fixed point, the value of δy at A is equal to zero.

For stationary I the first variation must vanish even if B is fixed, i.e. when the first two terms on the right of (5) are zero. Since δy is arbitrary the arguments of § 1.4 can then be used to show that the integral in (5) can vanish only if the Eulerian equation

$$\frac{\partial F}{\partial y} - \frac{d}{dx}\left(\frac{\partial F}{\partial y'}\right) = 0, \tag{6}$$

is satisfied. But this condition by itself is not sufficient in the present case, for when (6) is satisfied and B varies along Γ_2, (5) reduces to

$$\delta I_1 = F_b\, dx_b + \left(\frac{\partial F}{\partial y'}\right)_b \delta y_b. \tag{7}$$

With the help of Fig. VIII. 1 the right-hand side of (7) can be put into a much more useful form. Through B draw the ordinate so as to cut the varied curve at N and the line through B parallel to the x-axis at M. Then $\delta y_b = BM - NM$.

Now, for sufficiently small displacement BB', we have, from (2), $BM = g_2'(b)\, dx_b$, and from the equation of the varied curve, $NM = y'(b,\epsilon)\, dx_b$. But evidently $y'(b,\epsilon)\, dx_b$ differs from $y'(b,0)\, dx_b$ by second-order quantities only, so that

$$\delta y_b = (g_2' - y')_b\, dx_b, \tag{8}$$

where g_2' and y' are, respectively, the slopes of Γ_2 and the extremal at B. Hence (7) can be written in the form

$$\delta I_1 = \left\{F + (g_2' - y')\frac{\partial F}{\partial y'}\right\}_b dx_b. \tag{9}$$

Finally to make I stationary and $dI/d\epsilon$ vanish we must have, in addition to (6),

$$\left\{F + (g_2' - y')\frac{\partial F}{\partial y'}\right\}_b = 0. \tag{10}$$

This equation is known as a transversality condition and Γ_2 is said to be transversal to the extremal at B.

8.3. First variation of an integral with both end points variable

If A can be displaced along the curve Γ_1, whose equation is $y = g_1(x)$, it is evident that we must have another transversality condition analogous to (10), § 8.2. The complete result is summed up in the following theorem.

THEOREM 15. *If the end points A and B of the range of integration of the integral $I = \int\limits_a^b F(x,y,y')\,dx$ can be displaced along prescribed curves, then I is stationary when the following necessary conditions are satisfied:*

(i) *y, the ordinate of the extremal, satisfies the Eulerian equation*

$$\frac{\partial F}{\partial y} - \frac{d}{dx}\left(\frac{\partial F}{\partial y'}\right) = 0, \tag{1}$$

(ii) *at $x = a$*
$$F + (g_1' - y')\frac{\partial F}{\partial y'} = 0, \tag{2}$$

where a is the abscissa of the end point, A, which can be displaced along the curve $y = g_1(x)$,

(iii) *at $x = b$*
$$F + (g_2' - y')\frac{\partial F}{\partial y'} = 0, \tag{3}$$

where b is the abscissa of the end point, B, which can be displaced along the curve $y = g_2(x)$.

In these equations y' is the slope of the extremal and g_1', g_2' are respectively the slopes of the displacement curves of A and B at $x = a$ and $x = b$.

We shall mostly exclude the case when the extremal touches Γ_1 at A or Γ_2 at B, so that at $x = a$, $g_1' - y' \neq 0$ and at $x = b$, $g_2' - y' \neq 0$. From (2) and (3) this assumption is equivalent to the statement that $F(x,y,y')$ does not vanish at $x = a$ or at $x = b$.

8.4. Illustrations of the theory

EXAMPLE 1. The end points, A and B, of the arc of integration of the integral $I = \int_A^B G(x,y)(1+y'^2)^{\frac{1}{2}}\,dx$ can be displaced along given curves Γ_1 and Γ_2. If $G(x,y)$ does not vanish at A or B, prove that the extremal is orthogonal to Γ_1 and Γ_2.

From (2), § 8.3, we have at the point A

$$G(x,y)(1+y'^2)^{\frac{1}{2}}+(g_1'-y')G(x,y)\frac{y'}{(1+y'^2)^{\frac{1}{2}}} = 0. \tag{1}$$

This is easily simplified to $g_1'y'+1 = 0$, showing that the extremal and Γ_1 intersect orthogonally at A. Similarly for the intersection at B.

This result contains many well-known theorems as special cases and shows that transversality can in some ways be regarded as a generalization of orthogonality. Some simple illustrations are (i) the shortest distance from a point to a line is perpendicular to the line, (ii) the shortest distance between two non-coplanar lines lies along their common perpendicular, (iii) the shortest distance between two curves lies along a common normal.

The result is also applicable to dynamical problems, particularly when the principle of least action is employed. For a particle of unit mass moving in a conservative field of force the principle states (§ 5.6) that $2\int T\,dt = \int v(1+y'^2)^{\frac{1}{2}}\,dx$ is a minimum, where v is the velocity. Since v is a function of x and y, the conditions of example 1 are satisfied.

EXAMPLE 2. With the conditions of example 1 except that

$$I = \int_A^B G(x,y)(1+y'^2)^{\frac{1}{2}}e^{-(\tan^{-1}y')\cot\alpha}\,dx,$$

prove that the extremal intersects Γ_1 and Γ_2 at an angle α.

8.5. The Brachistochrone

The problem of § 1.11 can be generalized in the following manner.

Given a vertical plane containing a straight line L and a point A not on L. A particle slides along a smooth curve in the plane starting from rest at A and terminating at a point of L. Find the curve for which the time from A to L is a minimum.

Take A as the origin, the horizontal line through A in the plane as the x-axis, and the downward vertical through A as the

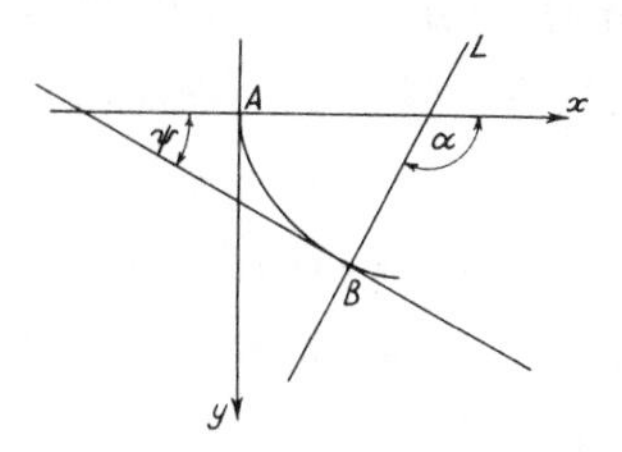

FIG. VIII. 2.

positive y-axis, Fig. VIII. 2. The velocity at depth y is $(2gy)^{\frac{1}{2}}$ and if the trajectory meets L at the point B, then the time from A to B is

$$\int_A^B \frac{ds}{v} = \frac{1}{(2g)^{\frac{1}{2}}} \int_A^B \frac{1}{y^{\frac{1}{2}}} (1+y'^2)^{\frac{1}{2}}\, dx. \tag{1}$$

The Eulerian equation, (1), § 8.3, for this integral has already been considered in § 1.11 and we shall quote the results obtained in that section. By using theorem 2, § 1.4, the characteristic equation can be integrated and simplified to

$$y(1+y'^2) = 2c, \tag{2}$$

where c is an arbitrary constant. On writing $y' = \tan\psi$, so that ψ is the angle between the x-axis and the tangent to the extremal, we have

$$y = c(1+\cos 2\psi). \tag{3}$$

Since

$$x = \int \cot\psi\, dy$$

we also have

$$x = b-c(2\psi+\sin 2\psi), \tag{4}$$

where b is a second arbitrary constant.

Equations (3) and (4) show that the curve is a cycloid whose directrix is the x-axis and the radius of whose generating circle is c.

We now proceed to evaluate the constants b and c from the given end conditions. On substituting $(0, 0)$, the coordinates of A, in (3) and (4) we get $\psi = \frac{1}{2}\pi$ and $b = \pi c$. Equation (4) then becomes

$$x = c(\pi-2\psi-\sin 2\psi), \tag{5}$$

showing that the cycloid has a cusp at A. At the point B example 1, § 8.4, shows that the cycloid is perpendicular to the line L. Let

$$y = (x-a)\tan\alpha \tag{6}$$

be the equation of the line L (Fig. VIII. 2), where a and α are given constants. Since $\tan\psi$ is the slope of the tangent at B we must have

$$\psi = \alpha - \tfrac{1}{2}\pi, \tag{7}$$

so that from (3) and (5) the coordinates of B are

$$[c(2\pi-2\alpha+\sin 2\alpha),\quad c(1-\cos 2\alpha)].$$

Since B is on (6) we have

$$c(1-\cos 2\alpha) = [c(2\pi-2\alpha+\sin 2\alpha)-a]\tan\alpha,$$

which simplifies to

$$c = \frac{a}{2(\pi-\alpha)}. \tag{8}$$

We thus determine c, the radius of the generating circle, and by substitution in (3) and (5) obtain the extremal required.

8.6. The second variation

As in § 8.2, we commence by keeping the end-point A fixed and allowing B to vary along the curve Γ_2. The result is easily generalized to allow for the displacement of A along the curve Γ_1.

On varying (4), § 8.2, from B to B' and differencing we have

$$d^2I = \left\{F\,d^2x+\frac{dF}{dx}(dx)^2+2\frac{\partial F}{\partial y}\,\delta y dx+2\frac{\partial F}{\partial y'}\,\delta y' dx\right\}_b+$$
$$+\int\limits_a^b \left\{(\delta y)^2\frac{\partial^2 F}{\partial y^2}+2\,\delta y\delta y'\frac{\partial^2 F}{\partial y\partial y'}+(\delta y')^2\frac{\partial^2 F}{\partial y'^2}+\right.$$
$$\left.+\frac{\partial F}{\partial y}\,\delta^2 y+\frac{\partial F}{\partial y'}\,\delta^2 y'\right\}dx. \tag{1}$$

The subscript b and the symbol δy have the meanings assigned to them in § 8.2 and the interpretation of d^2x will be obtained shortly, (5) below. The other terms are obtained as follows.

The terms $\dfrac{\partial F}{\partial y}\,\delta y dx+\dfrac{\partial F}{\partial y'}\,\delta y' dx$ occur once owing to the variation of $F_b\,dx_b$ from B to N, Fig. VIII. 1, and again on differentiating the integral of (4), § 8.2, and allowing for the change in the upper limit of integration b.

The term $\dfrac{dF}{dx}(dx)^2$ occurs owing to the variation of $F_b\,dx_b$ from N to B', Fig. VIII. 1.

The integrand of (1) is obtained by differentiating the integrand of (4), § 8.2, with respect to ϵ.

On integrating the last term of (1) by parts and using the characteristic equation (6), § 8.2, it follows that the last two terms of (1) reduce to $\left(\dfrac{\partial F}{\partial y'}\,\delta^2 y\right)_b$. We assume this in (4) below.

Again we may write (8), § 8.2, in the form

$$dy = y'\,dx+\delta y, \quad \text{at} \quad x = b, \tag{2}$$

where dy denotes the change in the ordinate of B when it is displaced along the curve Γ_2 to B'. From (2) we deduce

$$d^2y = y'\,d^2x+y''(dx)^2+2\,\delta y' dx+\delta^2 y, \quad \text{at} \quad x = b. \tag{3}$$

On using these two equations to eliminate δy and $\delta y'$ from (1) we obtain

$$d^2 I$$

$$= \left[F\,d^2x+(d^2y-y'\,d^2x)\frac{\partial F}{\partial y'}+\left(\frac{\partial F}{\partial x}-y'\,\frac{\partial F}{\partial y}\right)dx^2+2\frac{\partial F}{\partial y}\,dydx\right]_b+$$

$$+\int\limits_a^b \left\{(\delta y)^2\,\frac{\partial^2 F}{\partial y^2}+2\,\delta y \delta y'\,\frac{\partial^2 F}{\partial y \partial y'}+(\delta y')^2\,\frac{\partial^2 F}{\partial y'^2}\right\} dx. \tag{4}$$

Suppose now that the coordinates of points on the curve Γ_2 are given in terms of a parameter† t, i.e. on Γ_2 we have $x = \phi(t)$ and $y = \psi(t)$. Denoting differentiation with respect to t by a dot we have

$$dx = \dot{x}\,dt, \qquad d^2x = \ddot{x}\,dt^2+\dot{x}\,d^2t$$

$$dy = \dot{y}\,dt, \qquad d^2y = \ddot{y}\,dt^2+\dot{y}\,d^2t. \tag{5}$$

On inserting these values in (4) and using the transversality condition (10), § 8.2, it is found, since $g_2' = \dot{y}/\dot{x}$, that the terms in d^2t cancel. We are now able to divide (4) by $(d\epsilon)^2$ and make $d\epsilon$

† No conception of time is associated with the parameter t.

tend to zero and then place $\epsilon = 0$. The ordinate of the extremal is $y(x, 0)$ and that of the varied curve is $y(x, \epsilon)$, so that

$$\delta y = \frac{\partial}{\partial \epsilon} y(x, \epsilon)\, d\epsilon.$$

For brevity we shall write

$$\lim_{d\epsilon \to 0} \frac{\delta y}{d\epsilon} = \eta, \quad \lim_{d\epsilon \to 0} \frac{\delta y'}{d\epsilon} = \eta', \text{ when } \epsilon = 0.$$

We then deduce from (4) that

$$\left(\frac{d^2 I}{d\epsilon^2}\right)_{\epsilon=0} = \left[F\ddot{x} + (\ddot{y} - y'\ddot{x})\frac{\partial F}{\partial y'} + \left(\frac{\partial F}{\partial x} - y'\frac{\partial F}{\partial y}\right)\dot{x}^2 + 2\frac{\partial F}{\partial y}\dot{x}\dot{y}\right]\left(\frac{dt}{d\epsilon}\right)^2_{\substack{x=b\\ \epsilon=0}} + \int_a^b \left\{\eta^2 \frac{\partial^2 F}{\partial y^2} + 2\eta\eta' \frac{\partial^2 F}{\partial y \partial y'} + \eta'^2 \frac{\partial^2 F}{\partial y'^2}\right\} dx. \quad (6)$$

Here $\epsilon = 0$ in all the terms of the equation, including those inside the integral sign and $x = b$ in all terms of the expression inside the square brackets.

It is now easy to allow for the additional variation of the other end-point A along the curve Γ_1. If the coordinates of points on Γ_1 are expressed as functions of a parameter λ, then we must subtract from (6) a term analogous to that in square brackets but with $\dot{x}$ replaced by $dx/d\lambda$, etc., $dt/d\epsilon$ replaced by $d\lambda/d\epsilon$, and with values taken at $x = a$, $\epsilon = 0$. We shall denote the complete result by I_2, the terms in the square bracket of (6) by β, and the corresponding terms which arise for the variation of A by α. Further, we shall write $\partial^2 F/\partial y^2 = F_{yy}$, $\partial^2 F/\partial y \partial y' = F_{yy'}$ and $\partial^2 F/\partial y'^2 = F_{y'y'}$, all taken at $\epsilon = 0$. Our final result is then

$$I_2 = \beta\left(\frac{dt}{d\epsilon}\right)^2_{\substack{x=b\\ \epsilon=0}} - \alpha\left(\frac{d\lambda}{d\epsilon}\right)^2_{\substack{x=a\\ \epsilon=0}} + \int_a^b \{\eta^2 F_{yy} + 2\eta\eta' F_{yy'} + \eta'^2 F_{y'y'}\}\, dx. \quad (7)$$

For maximum or minimum I the quantity I_2 must maintain a constant sign for all permissible variations. Our next aim is to establish a set of conditions sufficient to ensure such constancy of sign. We can, with great advantage, use once again the ideas of §§ 2.2, 2.3, and 2.4.

8.7. The accessory equation

When the Eulerian characteristic equation (1), § 8.3, has been solved F_{yy}, $F_{yy'}$, and $F_{y'y'}$ become known functions of x. We can then define the accessory equation (Jacobi's or the subsidiary characteristic equation) as in § 2.4, namely

$$\left(F_{yy}-\frac{dF_{yy'}}{dx}\right)u-\frac{d}{dx}\left(F_{y'y'}\frac{du}{dx}\right) = 0. \tag{1}$$

This is a second-order differential equation for u so that the solution contains two arbitrary constants. It is the characteristic equation for the integral of (7), § 8.6, if we take x to be the independent and η to be the dependent variable. Let

$$y = s(x, c_1, c_2),$$

where c_1 and c_2 are arbitrary constants, be the solution of (1), § 8.3. Then as in § 2.8 we can prove that $\partial y/\partial c_1$ and $\partial y/\partial c_2$ are solutions of (1).

With the help of (1) and some simple, but lengthy, transformations we can express I_2 in a form whose sign is easily determined. Writing

$$2\Omega = \eta^2 F_{yy}+2\eta\eta' F_{yy'}+\eta'^2 F_{y'y'} \tag{2}$$

and observing that Ω is homogeneous in η and η' of degree two, we have

$$\int 2\Omega\,dx = \int \left(\eta\frac{\partial\Omega}{\partial\eta}+\eta'\frac{\partial\Omega}{\partial\eta'}\right)dx \tag{3}$$

$$= \eta\frac{\partial\Omega}{\partial\eta'}+\int \eta\left\{\frac{\partial\Omega}{\partial\eta}-\frac{d}{dx}\left(\frac{\partial\Omega}{\partial\eta'}\right)\right\}dx \tag{4}$$

on integration by parts. Hence, from (2), we have

$$\int 2\Omega\,dx = \eta\frac{\partial\Omega}{\partial\eta'}+\int \eta\left\{(\eta F_{yy}+\eta' F_{yy'})-\frac{d}{dx}(\eta F_{yy'}+\eta' F_{y'y'})\right\}dx$$

$$= \eta\frac{\partial\Omega}{\partial\eta'}+\int \eta\left\{\left(F_{yy}-\frac{dF_{yy'}}{dx}\right)\eta-\frac{d}{dx}(\eta' F_{y'y'})\right\}dx. \tag{5}$$

Now if u is a solution of (1) and u' is its derivative we may re-write (5) in the form

$$\int 2\Omega\,dx = \eta\frac{\partial\Omega}{\partial\eta'}+\int\frac{\eta}{u}\left\{\eta\frac{d}{dx}(u'F_{y'y'})-u\frac{d}{dx}(\eta'F_{y'y'})\right\}dx.$$

After integration by parts we have

$$\int 2\Omega\,dx = \eta\frac{\partial\Omega}{\partial\eta'}+\left(\eta^2\frac{u'}{u}-\eta\eta'\right)F_{y'y'}-\int F_{y'y'}\left\{u'\frac{d}{dx}\left(\frac{\eta^2}{u}\right)-\eta'^2\right\}dx.$$

This, in turn, reduces to

$$\int 2\Omega\,dx = \frac{\eta^2}{u}(uF_{yy'}+u'F_{y'y'})+\int F_{y'y'}\left(\eta'-\frac{u'}{u}\eta\right)^2dx, \tag{6}$$

where the terms not inside the integral have values corresponding to the end points of the arc of integration. The quantity η has been defined in § 8.6 as equal to $\lim\limits_{d\epsilon\to 0}\frac{\delta y}{d\epsilon}$, when $\epsilon = 0$. Hence from (8), § 8.2 it follows that

$$\left.\begin{aligned}&\text{at } x = a \qquad && \eta = (g_1'-y')\frac{dx}{d\lambda}\frac{d\lambda}{d\epsilon}\\ &\text{and at } x = b \qquad && \eta = (g_2'-y')\dot{x}\frac{dt}{d\epsilon}.\end{aligned}\right\} \tag{7}$$

We now choose two solutions of the accessory equation (1), $u_1(x)$ and $u_2(x)$. For brevity these will be denoted by u_1 and u_2 and their derivatives by u_1' and u_2' respectively. Since (1) is a second-order differential equation we may choose each of these solutions to satisfy the following two conditions:

$$\left.\begin{aligned}&\text{at } A, \quad && x = a,\ u_1 = (g_1'-y')\frac{dx}{d\lambda};\\ & && (u_1F_{yy'}+u_1'F_{y'y'})(g_1'-y')\frac{dx}{d\lambda} = -\alpha\\ &\text{at } B, \quad && x = b,\ u_2 = (g_2'-y')\dot{x};\\ & && (u_2F_{yy'}+u_2'F_{y'y'})(g_2'-y')\dot{x} = -\beta,\end{aligned}\right\} \tag{8}$$

where α and β are the quantities employed in (7), § 8.6.

In (6) take $u = u_1$, $a \leqslant x < c$, $u = u_2$, $c < x \leqslant b$. We obtain

$$\int\limits_a^b 2\Omega\, dx = \alpha\left(\frac{d\lambda}{d\epsilon}\right)^2 - \beta\left(\frac{dt}{d\epsilon}\right)^2 + \left\{\frac{\eta^2}{u_1 u_2}(u_1' u_2 - u_1 u_2')F_{y'y'}\right\}_{x=c} +$$
$$+ \int\limits_a^c F_{y'y'}\left(\eta' - \frac{u_1'}{u_1}\eta\right)^2 dx + \int\limits_c^b F_{y'y'}\left(\eta' - \frac{u_2'}{u_2}\eta\right)^2 dx. \quad (9)$$

In all terms $\epsilon = 0$ and inside the brackets $\{\,\}$ $x = c$, as indicated by the subscript. Combining (7), § 8.6, with (2) and (9) we finally obtain

$$I_2 = \left\{\frac{\eta^2}{u_1 u_2}(u_1' u_2 - u_1 u_2')F_{y'y'}\right\}_{x=c} + \int\limits_a^c F_{y'y'}\left(\eta' - \frac{u_1'}{u_1}\eta\right)^2 dx +$$
$$+ \int\limits_c^b F_{y'y'}\left(\eta' - \frac{u_2'}{u_2}\eta\right)^2 dx. \quad (10)$$

In this expression η is any permissible variation satisfying the end conditions (8), u_1 and u_2 are solutions of the accessory equation (1), and, as indicated by the subscript, in terms outside the integral signs, we must have $x = c$, $a \leqslant c \leqslant b$.

The accessory equation (1) is a special case of a type of differential equation known as Sturm–Liouville equations. Some properties of the solutions of these equations have been obtained in § 2.18 and one important property is that $(u_1' u_2 - u_1 u_2')F_{y'y'}$ is constant.

8.8. Focal points

Consider $u_1(x)$, that solution of the accessory equation (1), § 8.7, which satisfies the end conditions (8), § 8.7, at the point $x = a$. A point P on the extremal is said to be a focal point of Γ_1 when its abscissa is a zero of $u_1(x)$. Similarly for focal points of Γ_2. These are points of the extremal whose abscissae are zeros of $u_2(x)$, where $u_2(x)$ is that solution of (1), § 8.7, which satisfies the end conditions (8), § 8.7, at the point B, $x = b$.

Focal points play in variable end point theory the part played by conjugate points in fixed end point theory, see § 2.6. If one of the end points, e.g. A, is fixed the end conditions at $x = a$, (8), § 8.7, are replaced by $u_1(a) = 0$ and the focal points for Γ_1 then become the conjugate points of A.

Suppose that there exists a focal point of Γ_1 at $x = c$, where $a \leqslant c \leqslant b$. Then $u_1(c) = 0$ and so a permissible choice for η is

$$\eta = u_1(x), \qquad a \leqslant x \leqslant c, \qquad \eta = 0, \qquad c \leqslant x \leqslant b.$$

For the necessary end condition at A, (7), § 8.7, is satisfied and the other end B can be kept fixed. From (10), § 8.7, it follows that $I_2 = 0$ and therefore that when $I = \int\limits_a^b F(x,y,y')\,dx$ is stationary its first and second variations vanish. For such a case the value of δI will then depend upon the third variation which, for sufficiently small ϵ, contains terms of the type ϵ^3. Consequently δI cannot in general maintain the constant sign required at a maximum or minimum value of I. To obtain a maximum or minimum value of I we must therefore exclude all focal points from the arc AB of the extremal.

If one end point, e.g. A, is fixed, these arguments show that the conjugate points of A must be excluded from the arc AB of the extremal as in § 2.6.

The results obtained so far can be summed up in the following theorem:

THEOREM 16. *If the end points of the arc of integration of* $I = \int\limits_a^b F(x,y,y')\,dx$ *can be displaced along prescribed curves, then to ensure that I has a maximum or minimum value the following conditions must be satisfied in addition to those of Theorem* 15, § 8.3:

(iv) *No focal points of either of the prescribed curves* Γ_1, Γ_2 *exist within or at either end of the extremal arc AB.*

(v) *If* $u_1(x)$ *and* $u_2(x)$ *are solutions of the accessory equation* (1), § 8.7, *satisfying the end conditions* (8), § 8.7, *then*

$$\{u_1(x)u_2(x) - u_1(x)u_2(x)\}$$

has the same sign as $u_1(x)u_2(x)$ *throughout the range of integration AB.*

(vi) $\partial^2 F/\partial y'^2$ $(= F_{y'y'})$ *maintains constant sign throughout this range.*

(vii) *The Bliss condition. This deals with the case when the focal points lie outside the arc AB and on the same side of it, as in Figs. VIII. 3 and VIII. 4, where S_1 and S_2 are, respectively, the focal points of Γ_1 and Γ_2. On traversing the extremal in the direction from A to B the condition requires that the order of the points A, B, S_1, S_2, should be either ABS_2S_1, as in Fig. VIII. 3, or S_2S_1AB, as in Fig. VIII. 4. Possibilities such as ABS_1S_2 and S_1S_2AB must be excluded.*

With these conditions fulfilled I has a maximum when $F_{y'y'}$ is negative and a minimum when it is positive.

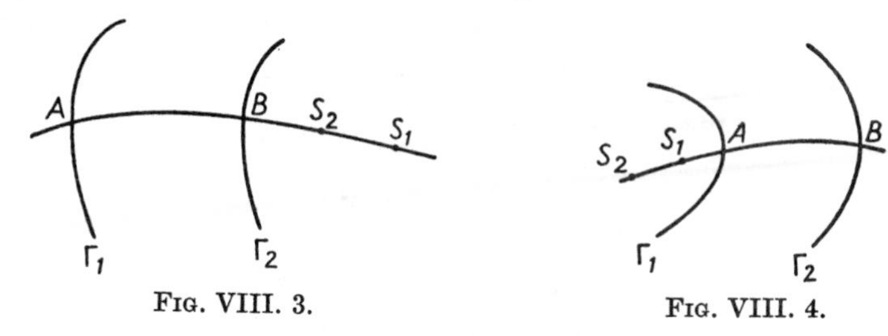

FIG. VIII. 3. FIG. VIII. 4.

The proof of the necessity of the Bliss condition will be postponed to § 8.14. The necessity of the other conditions is evident from the form of (10), § 8.7. For the exclusion of the focal points together with (v) and (vi) ensure that I_2 is positive definite if $F_{y'y'}$ is positive, and negative definite if it is negative.

8.9. The determination of focal points, (i) Geometrical

The method of envelopes, used in § 2.9 for conjugate points, can be applied with equal facility to the problem of finding the positions of focal points.

Since the Eulerian characteristic equation (1), § 8.3 is of the second order, it follows that the extremals form a two-parameter family of curves with equations of the form

$$y = s(x, c_1, c_2), \tag{1}$$

where c_1 and c_2 are arbitrary constants. Suppose that those members of (1) which satisfy one of the transversality conditions, say (2), § 8.3, possess an envelope. Then we can prove that the focal points of Γ_2, B's displacement curve, are at the points of contact of the extremal and the envelope.

To prove this let a neighbouring curve to (1) have the equation

$$y = s(x, c_1+\delta c_1, c_2+\delta c_2) \tag{2}$$

$$= s(x, c_1, c_2)+\frac{\partial y}{\partial c_1}\delta c_1+\frac{\partial y}{\partial c_2}\delta c_2, \tag{3}$$

where terms containing δc_1^2, $\delta c_1 \delta c_2$, δc_2^2 and higher powers are neglected.

Since $\frac{\partial y}{\partial c_1}\delta c_1+\frac{\partial y}{\partial c_2}\delta c_2$ is the difference between the ordinates of two neighbouring curves it may be denoted by δy, defined in § 8.2. Again, since $\partial y/\partial c_1$ and $\partial y/\partial c_2$ both satisfy the accessory equation (1), § 8.7, see § 2.8 for proof, then δy must also satisfy (1), § 8.7. We may then write

$$\frac{\partial y}{\partial c_1}\delta c_1+\frac{\partial y}{\partial c_2}\delta c_2 = \delta y = u_2\,dt = u_2\frac{dt}{d\epsilon}\delta\epsilon, \tag{4}$$

where (i) t is the parameter in terms of which the coordinates of points on Γ_2 are expressed, (ii) ϵ is as defined in § 8.2, and (iii) u_2 is a solution of (1), § 8.7.

Evidently the points of intersection of (1) and (2) occur at the zeros of u_2. But a solution of (1), § 8.7, which satisfies the conditions (8), § 8.7, at $x = b$ has its zeros at the focal points of Γ_2. Hence, if we can prove that u_2 satisfies these conditions, then (1) and (2) must, in the limit, intersect at the focal points of Γ_2.

To prove that u_2 satisfies the first of these conditions we combine (4) above and (8), § 8.2, when it follows immediately that

$$u_2 = (g_2'-y')\dot{x} \tag{5}$$

at $x = b$. Hence u_2 satisfies the first condition.

To prove that u_2 satisfies the second of the conditions,

$$(u_2 F_{yy'}+u_2' F_{y'y'})u_2 = -\beta \tag{6}$$

at $x = b$, is much more difficult. The method of proof is as follows: Since Γ_2 is transversal to (1) at B, where $x = b$, and transversal to (2) at B', Fig. VIII. 1, we have two transversality conditions,

$$F+(g_2'-y')\frac{\partial F}{\partial y'} = 0 \tag{7}$$

at B and a corresponding one at B'. On subtracting one of these equations from the other, dividing by dt, and proceeding to the limit $dt \to 0$, we obtain (6). This procedure is evidently equivalent to differentiating the left-hand side of (7). Or, on eliminating $(g_2'-y')$ by means of (4) and (5), we may vary $F\,dx+\delta y\dfrac{\partial F}{\partial y'}$ from B to B' and then insert $x = b$.

This variation, which arises from the displacement of B to B', can be performed in two stages. The first stage is due to the displacement from B to N, Fig. VIII. 1, and here we must allow for the change in the ordinate from y to $y+\delta y$ together with the consequential changes in $\delta y'$. The second stage is due to the displacement from N to B', Fig. VIII. 1, and here we allow for the change from x to $x+\delta x$. The results are tabulated below, where the expression to be varied is in the first column, the variation in the first stage is in the second column, and that in the second stage is in the third column.

$F\,dx$	$\dfrac{\partial F}{\partial y}\delta y dx+\dfrac{\partial F}{\partial y'}\delta y' dx$	$F\,d^2x+\dfrac{dF}{dx}(dx)^2$
δy	$\delta^2 y$	$\delta y' dx$
$\dfrac{\partial F}{\partial y'}$	$\dfrac{\partial^2 F}{\partial y \partial y'}\delta y+\dfrac{\partial^2 F}{\partial y'^2}\delta y'$	$\dfrac{d}{dx}\left(\dfrac{\partial F}{\partial y'}\right)dx$

From the characteristic equation we can write $\dfrac{d}{dx}\left(\dfrac{\partial F}{\partial y'}\right) = \dfrac{\partial F}{\partial y}$ in the last term of the third row.

On using this table the result of varying

$$F\,dx+\delta y\frac{\partial F}{\partial y}$$

is to give us a number of terms which we classify into three groups (*a*), (*b*), and (*c*). Group (*a*) contains all the terms inside the brackets $\{\ \}_b$ of (1), § 8.6, group (*b*) consists of the term

$\dfrac{\partial F}{\partial y'}\delta^2 y$, which arises from (1), § 8.6, from integration by parts, and group (c) consists of the terms

$$\delta y\left(\frac{\partial^2 F}{\partial y \partial y'}\delta y+\frac{\partial^2 F}{\partial y'^2}\delta y'\right).$$

On dividing by dt^2 and proceeding to the limit, from (4), the terms of group (c) lead to the left-hand side of (6) and, from the analysis of § 8.6, the terms of groups (a) and (b) sum to the expression denoted there by β. Hence u_2 satisfies conditions (8), § 8.7, at $x = b$, and the focal points of Γ_2 must be at the limiting points of intersection of (1) and (2). The envelope property stated at the commencement of this section then immediately follows.

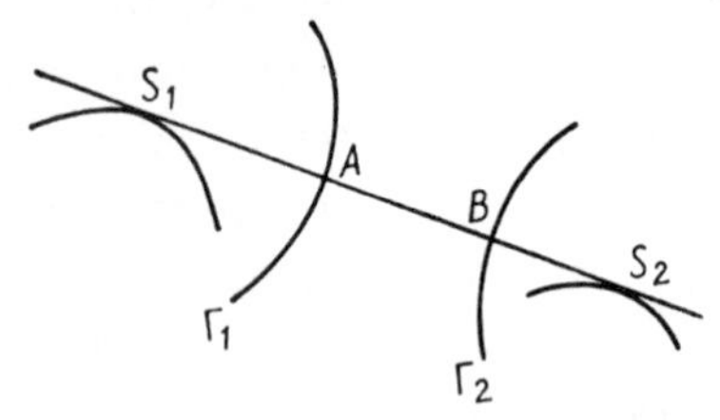

FIG. VIII. 5.

An illustration is given by the problem of finding the shortest distance between two coplanar curves Γ_1 and Γ_2. The integral to be minimized is $\int (1+y'^2)^{\frac{1}{2}}\,dx$ and the extremals are members of the family of straight lines $y = c_1 x+c_2$. Example 1, § 8.4, shows that when the extremals satisfy the transversality conditions they are normal to Γ_1 and Γ_2.

Consider now the family of straight lines normal to Γ_1. They all touch the evolute of Γ_1. Consequently the focal point of Γ_1 for an extremal intersecting it at A is S_1, the centre of curvature of A for Γ_1 (Fig. VIII. 5). Let AB be a common normal to Γ_1 and Γ_2 with S_2 the centre of curvature of Γ_2 at B, Fig. VIII. 5. Then AB is the required minimum distance if (i) S_1 and S_2 lie outside the segment AB and (ii) the Bliss condition of Theorem 16, § 8.8, is satisfied.

This example illustrates the fact that the position of the focal points depends upon the curvature of the transversals. It has been shown† that as the focal point of Γ_1 varies from A to the first conjugate of A, then the radius of curvature of Γ_1 at A varies from $-\infty$ to $+\infty$.

8.10. The determination of focal points, (ii) Analytical

The abscissae of focal points are the zeros of a function $u(x)$ which satisfies (i) the accessory equation (1), § 8.7, and (ii) one of the pair of end conditions (8), § 8.7. We shall illustrate this by means of a catenary problem similar to the ones discussed in §§ 1.7, 2.11, and 2.13, but with one end point variable.

A uniform flexible heavy string has one end coiled and lying on a horizontal plane and the other end B attached to a small smooth ring of negligible mass. The string rises vertically from the coil and passes over a fixed small smooth peg A before reaching B. If the ring can slide on a fixed straight line situated in a vertical plane through A, find the equilibrium position of the string.

The potential energy of the arc AB is a constant multiple of

$$I = \int\limits_A^B y(1+y'^2)^{\frac{1}{2}}\,dx = \int\limits_A^B F\,dx, \tag{1}$$

where the positive value of the root is taken. When the string is in stable equilibrium this integral must be a minimum.

The characteristic equation for (1) has already been obtained in § 1.7, where it is shown that the extremals are catenaries whose equations are

$$y = c\cosh\left(\frac{x+d}{c}\right), \tag{2}$$

where c and d are arbitrary constants. Example 1, § 8.4, shows that the transversality condition is satisfied if the given straight line is normal to the catenary (2) which passes through A. The coordinates of A and the equation of this line then suffice to determine the values of c and d.

† G. A. Bliss, *Lectures on the Calculus of Variations*, University of Chicago Press, § 66; O. Bolza, *Vorlesungen über Variationsrechnung*, Teubner, § 39.

From § 2.8 it follows that the solutions of the accessory equation (1), § 8.7, are

$$\frac{\partial y}{\partial d} = \sinh\left(\frac{x+d}{c}\right) \tag{3}$$

and

$$\frac{\partial y}{\partial c} = \cosh\left(\frac{x+d}{c}\right) - \frac{x+d}{c}\sinh\left(\frac{x+d}{c}\right). \tag{4}$$

We may therefore define $u_2(x)$, a general solution of the accessory equation, as follows:

$$u_2(x) = p\sinh\left(\frac{x+d}{c}\right) + q\left\{\cosh\left(\frac{x+d}{c}\right) - \frac{x+d}{c}\sinh\left(\frac{x+d}{c}\right)\right\}, \tag{5}$$

where p and q are arbitrary constants.

In this problem Γ_2 is the given straight line and for its parametric equations we take

$$x = a + t\cos\alpha; \qquad y = t\sin\alpha. \tag{6}$$

Evidently

$$\dot{x} = \cos\alpha, \qquad \dot{y} = \sin\alpha, \qquad \ddot{x} = \ddot{y} = 0. \tag{7}$$

From example 1, § 8.4, the transversality condition is satisfied if Γ_2 and the extremal intersect orthogonally at B. We can therefore draw up the following table of values at B:

$$\left.\begin{aligned} y' = -\cot\alpha, \quad y = c(1+y'^2)^{\frac{1}{2}} = c\,\mathrm{cosec}\,\alpha, \\ \frac{\partial F}{\partial x} = 0, \quad \frac{\partial F}{\partial y} = \mathrm{cosec}\,\alpha, \quad F_{yy'} = -\cos\alpha, \quad F_{y'y'} = c\sin^2\alpha. \end{aligned}\right\} \tag{8}$$

From § 8.6 we can determine β, which is equal to the expression inside the square brackets of (6), § 8.6. Using (7) and (8) it follows that

$$\begin{aligned} \beta &= \frac{\partial F}{\partial y}(2\dot{x}\dot{y} - y'\dot{x}^2) \\ &= \cos\alpha + \cos\alpha\,\mathrm{cosec}^2\alpha. \end{aligned} \tag{9}$$

Applying these results to the end conditions (8), § 8.7, we have

$$\begin{aligned} u_2(b) = (g_2' - y')\dot{x} &= (\tan\alpha + \cot\alpha)\cos\alpha \\ &= \mathrm{cosec}\,\alpha \end{aligned} \tag{10}$$

and

$$\{u_2(b)F_{yy'} + u_2'(b)F_{y'y'}\}(g_2' - y')\dot{x} = -\beta, \tag{11}$$

where g_2' is the slope of the given line. Solving these equations for $u_2'(b)$ we obtain, after some simplification,

$$u_2'(b) = -\frac{1}{c}\cot\alpha. \tag{12}$$

The constants p and q of (5) can now be calculated. By choosing a point underneath the vertex of the catenary as the origin we make $d = 0$ and simplify the calculations without loss of generality. We then have at the point $x = b$

$$p\sinh\left(\frac{b}{c}\right)+q\left\{\cosh\left(\frac{b}{c}\right)-\frac{b}{c}\sinh\frac{b}{c}\right\} = \operatorname{cosec}\alpha, \tag{13}$$

$$\frac{p}{c}\cosh\left(\frac{b}{c}\right)+\frac{q}{c}\left\{-\frac{b}{c}\cosh\left(\frac{b}{c}\right)\right\} = -\frac{1}{c}\cot\alpha. \tag{14}$$

On noting that at $x = b$ the slope of the catenary is $\sinh(b/c)$ it follows that $\sinh(b/c) = y' = -\cot\alpha$, $\cosh(b/c) = \operatorname{cosec}\alpha$. From (13) and (14) we then obtain

$$\left.\begin{aligned} q &= \sin^2\alpha, \\ p &= \sin^2\alpha\left\{\frac{b}{c}+\sinh\left(\frac{b}{c}\right)\cosh\left(\frac{b}{c}\right)\right\}. \end{aligned}\right\} \tag{15}$$

Finally from (5) the equation for abscissa of the focal points, $u_2(x) = 0$, becomes

$$\left\{\frac{b}{c}+\sinh\left(\frac{b}{c}\right)\cosh\left(\frac{b}{c}\right)\right\}\sinh\frac{x}{c}+\cosh\left(\frac{x}{c}\right)-\frac{x}{c}\sinh\left(\frac{x}{c}\right) = 0. \tag{16}$$

This equation can be given an interesting geometrical interpretation. If the normal at B to the catenary $y = c\cosh(x/c)$, Fig. VIII. 6, meets the directrix at G, and O is the origin, then

$$OG = b+c\sinh\left(\frac{b}{c}\right)\cosh\left(\frac{b}{c}\right). \tag{17}$$

If the tangent at S, a focal point of BG, meets the directrix at G_1, then

$$OG_1 = x-\frac{y}{(1+y'^2)^{\frac{1}{2}}}, \tag{18}$$

where x is the abscissa of S. From the equation of the catenary we obtain

$$OG_1 = x-c\frac{\cosh(x/c)}{\sinh(x/c)} = OG \tag{19}$$

from (16) and (17). Consequently the tangent at S, the normal at B and the directrix are concurrent at G. Since there are two tangents from G to the catenary there are two focal points.

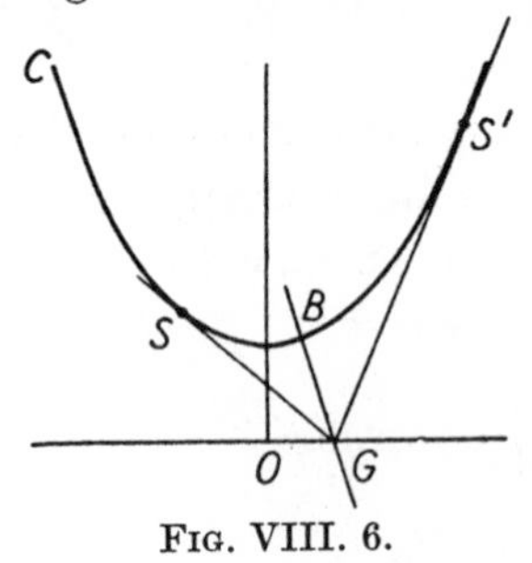

FIG. VIII. 6.

If C is conjugate to B, then the tangents at B and C intersect on the directrix, § 2.13, and so one of the focal points, S, lies between C and B and the other, S', lies outside the arc CB, see Fig. VIII. 6. The appropriate focal point of BG is S when A lies between C and B and S' when B lies between C and A.

On taking the positive value of the root it follows that

$$F_{y'y'} = y(1+y'^2)^{-\frac{3}{2}}$$

is positive at all points of the catenary. Hence the integral I of (1) has a minimum if A lies between S and B or between B and S' (Fig. VIII. 6).

8.11. Hilbert's integral

Important deductions can be made from the results obtained in § 8.2. We consider, as before,

$$I = \int\limits_A^B F(x,y,y')\,dx = \int\limits_A^B F\,dx, \tag{1}$$

where A is a fixed end point and B can be displaced along the curve Γ_2 whose equation is $y = g_2(x)$. If F satisfies the Eulerian equation (6), § 8.2, so that the curve AB is an extremal, then (9), § 8.2, states that

$$\delta I = \left\{F+(g_2'-y')\frac{\partial F}{\partial y'}\right\} dx_{x=b}, \tag{2}$$

where g_2' and y' are respectively the slopes of Γ_2 and the extremal at B, $x = b$. Terms containing ϵ^2 and higher powers have been neglected.

Equation (2) is easily generalized to the case when A is no longer fixed but can be displaced along the curve Γ_1 whose equation is $y = g_1(x)$. The result is

$$\delta I = \left\{F+(g_2'-y')\frac{\partial F}{\partial y'}\right\}dx_2-\left\{F+(g_1'-y')\frac{\partial F}{\partial y'}\right\}dx_1, \tag{3}$$

where the subscripts 1 and 2 denote, respectively, values taken at A, $x = a$, and B, $x = b$. If A is displaced to A', a neighbouring

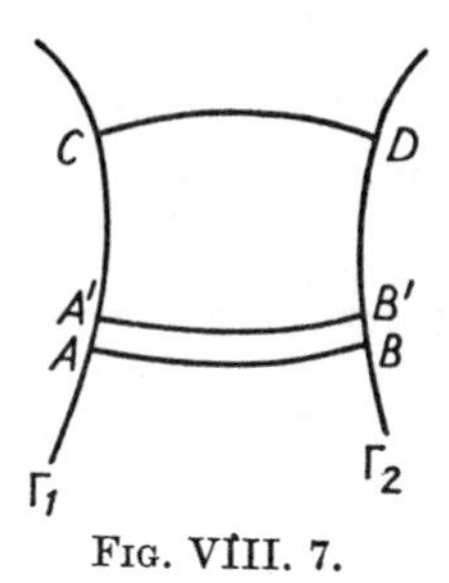

FIG. VIII. 7.

point on Γ_1, and B to B', a neighbouring point on Γ_2 then we also have (Fig. VIII. 7)

$$\delta I = \int\limits_{A'}^{B'} F(x,y,y')\,dx - \int\limits_{A}^{B} F(x,y,y')\,dx, \tag{4}$$

where the arcs AB and $A'B'$ are both extremals.

Consider now a displacement of the whole arc AB in which (i) A is displaced along Γ_1 to C and B along Γ_2 to D (Fig. VIII. 7), and (ii) the arc AB remains an extremal, i.e. its equation satisfies the Eulerian equation (6), § 8.2, in all intermediate positions. Then (3) holds throughout the displacement. We deduce that

$$\int\limits_{C}^{D} F(x,y,y')\,dx - \int\limits_{A}^{B} F(x,y,y')\,dx$$

$$= \int\limits_{B}^{D} \left\{F+(g_2'-y')\frac{\partial F}{\partial y'}\right\}dx_2 - \int\limits_{A}^{C} \left\{F+(g_1'-y')\frac{\partial F}{\partial y'}\right\}dx_1, \tag{5}$$

where the subscripts 1 and 2 denote integration along the curves Γ_1 and Γ_2 respectively. On writing

$$I_1^* = \int \left\{F+(g_1'-y')\frac{\partial F}{\partial y'}\right\}dx_1, \tag{6}$$

and similarly for I_2^*, (5) becomes

$$I(CD)-I(AB) = I_1^*(BD)-I_2^*(AC), \tag{7}$$

where the letters inside the brackets denote the end points of the range of integration in each case. The integral I^* is known as Hilbert's integral.

We can make C coincide with D and A with B without making the paths AC and BD coincide with each other. From (7) we then have $I^*(BD) = I^*(AC)$, which indicates that the value of I^* is in general, independent of the path of integration. This

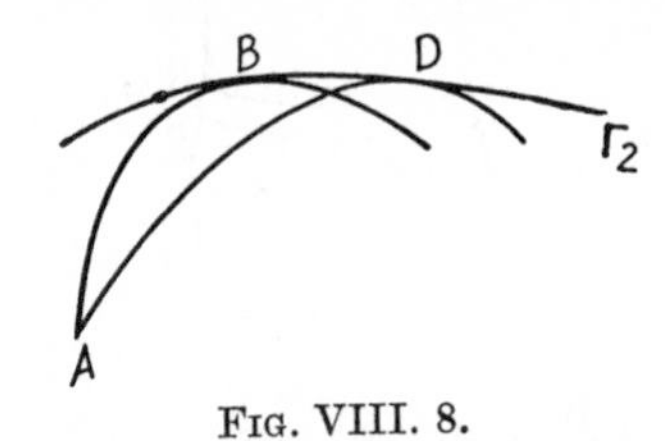

FIG. VIII. 8.

fundamental property of Hilbert's integral cannot be adequately discussed without the concept of a 'field' introduced in the next section. We shall therefore postpone the general discussion of I^* until § 8.13 and deal only with some special cases in this section.

Consider a one-parameter family of extremals passing through a fixed point A and possessing an envelope. In Fig. VIII. 8 AB and AD are two members of the family touching their envelope BD at B and D respectively. We can displace the arc AB to AD through a series of intermediate extremals each of which passes through A and touches the envelope BD. Such a displacement satisfies the requirements of (5) in the special case when A and C coincide. Now along the envelope we have $g_2' = y'$, so that in this case (5) reduces to

$$\int\limits_A^D F(x,y,y')\,dx = \int\limits_A^B F(x,y,y')\,dx + \int\limits_B^D F(x,y,y')\,dx, \tag{8}$$

where the first two integrals are taken along the extremals AD and AB respectively and the last along the envelope.

An example of a one-parameter family of extremals passing through a point occurs when a particle is projected under gravity from a given point A with given initial speed u. By varying the direction of projection we obtain different members of the family. This was discussed in §§ 1.8, 2.10, and 2.14 where, by means of the principle of least action, it was proved that the trajectories are extremals of the integral

$$\int (u^2-2gy)^{\frac{1}{2}}(1+y'^2)^{\frac{1}{2}}\,dx.$$

Restricting ourselves to one vertical plane through A, the extremals are found to be a family of parabolas passing through A, each touching another parabola having A as a focus. The axis of the enveloping parabola is the vertical through A and its latus rectum is $2u^2/g$. Equation (8), with

$$F(x,y,y') = (u^2-2gy)^{\frac{1}{2}}(1+y'^2)^{\frac{1}{2}},$$

holds for such a system.

Equation (8) can be generalized in a very interesting manner. If the extremals and their slopes y' are known at every point we can solve the first-order differential equation

$$F+(g'-y')\frac{\partial F}{\partial y'} = 0 \tag{9}$$

for g. A one-parameter family of curves is then obtained, one curve passing through each point of the plane. By the definition of § 8.2 the member of this family which passes through P is transversal to the extremal through P. If the arc AC of (5) is one of the curves of the family defined by (9), i.e. if we take $g_1' = g'$, then the last integral of (5) vanishes.

Consider now a one-parameter family of extremals possessing an envelope and let AB, CD be two of the members touching the envelope at B and D respectively. If in (5) the arc AB is displaced to CD so that Γ_1, the displacement curve of A, is one of the transversals defined by (9) and Γ_2 is the envelope BD on which $g_2' = y'$, see Fig. VIII. 9, then we obtain

$$\int\limits_C^D F(x,y,y')\,dx = \int\limits_A^B F(x,y,y')\,dx + \int\limits_B^D F(x,y,y')\,dx. \tag{10}$$

The first two of these integrals are taken along the extremals and the third along the envelope.

Equation (10) contains as a special case the well-known

'unwrapping' relation between a curve and its evolute. Let $F(x,y,y') = (1+y'^2)^{\frac{1}{2}}$ so that $I = \int (1+y'^2)^{\frac{1}{2}}\,dx = \int ds$ and the extremals are therefore straight lines. Equation (9) reduces to $g'y'+1 = 0$, showing that the lines are normal to Γ_1. Hence

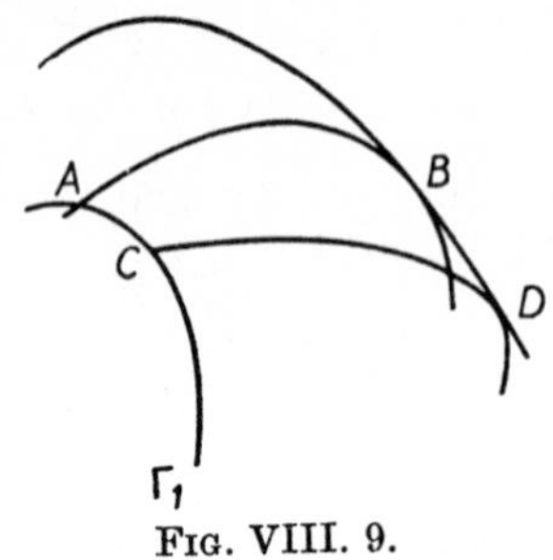

FIG. VIII. 9.

during the displacement from AB to CD we have a series of lines normal to the arc AC and so touching the evolute of AC. Equation (10) then becomes

$$CD = AB+\text{arc}\,BD, \tag{11}$$

a relation which expresses the well-known fact that a curve can be obtained by unwrapping a string from its evolute (Fig. VIII. 10).

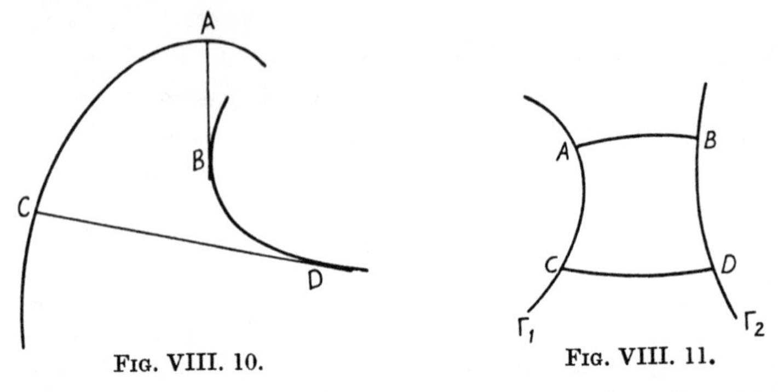

FIG. VIII. 10.

FIG. VIII. 11.

In the case when both Γ_1 and Γ_2 are transversals and so members of the family defined by (9), both the right-hand integrals of (5) vanish. We then have

$$\int\limits_A^B F(x,y,y')\,dx = \int\limits_C^D F(x,y,y')\,dx, \tag{12}$$

where the integrals are taken along the extremals AB and CD (Fig. VIII. 11).

8.12. Fields of extremals

The results of the previous section lead to the concept of a field of extremals which is much used in modern theory.

Starting with the integral

$$I = \int F(x, y, y')\, dx \tag{1}$$

we can, by means of the Eulerian characteristic equation, obtain a two-parameter family of extremals. When the Eulerian equation is solved we can then find a direction g' defined by the equation

$$F + (g' - y')\frac{\partial F}{\partial y'} = 0. \tag{2}$$

Suppose that by introducing some suitable restrictions we obtain from the two-parameter family of extremals a one-parameter family. Then a field of extremals is defined if no two members of this one-parameter family have a point in common. Through every point P of such a field there passes a unique extremal E and, if y' is its slope at P, we obtain from (2) a unique slope g' transversal to E. On solving (2) for the function g we obtain a one-parameter family of curves no two of which have a common point in the field. These are transversal to the extremals. The part of the plane to which the field is limited can evidently be covered by a network of extremals and transversals in such a manner that through each point of the field we can draw one extremal and one transversal only.

Hilbert's integral, as defined in § 8.11, is

$$I^* = \int\limits_A^B \left\{F + (g' - y')\frac{\partial F}{\partial y'}\right\} dx, \tag{3}$$

where g' is the slope of the arc of integration and y' is the slope of the extremal. If F is one-valued, or if multi-valued is restricted by the field to the values of one branch only, then the integrand of (3) is uniquely defined at every point of the path of integration and the integral can therefore be evaluated. The fundamental property of I^* is that the value of $I^*(AB)$ is independent of the path AB if the arc of integration lies entirely inside a field of extremals. We shall prove this in the next section.

If AB, the arc of integration, is an extremal, then $g' = y'$ and so

$$I^* = I. \tag{4}$$

If the arc AB is a transversal then, from (2), we have

$$I^* = 0. \tag{5}$$

The simplest example of a field is derived from the integral

$$I = \int F(y')\,dx. \tag{6}$$

The extremals are the lines $y = mx+n$, where m and n are arbitrary constants. We can obtain a field by confining ourselves to a family of lines parallel to a given direction. The transversals would then be the family of parallel lines perpendicular to them.

A field can be obtained for (6) in another way. Consider a family of lines concurrent at O. Then if we exclude a domain which contains the point O, the remaining segments of the extremals form a field. The transversals are arcs of concentric circles having O as a common centre.

A more complicated example is given by the integral

$$I = \int y(1+y'^2)^{\frac{1}{2}}\,dx. \tag{7}$$

The extremals are the catenaries $y = c\cosh\{(x+a)/c\}$, where a and c are arbitrary constants and the x-axis is a common directrix. From this two-parameter family we can obtain a one-parameter family by restricting ourselves to those curves which pass through a given point A on the positive y-axis. This sub-family has an envelope E_c (see Fig. VIII. 12), and if B lies outside E_c there are no members of the family which pass through both A and B. But if B lies inside E_c, i.e. in the region which contains the positive y-axis, then there are two catenaries of the sub-family passing through A and B. In Fig. VIII. 12 the two possible catenaries are shown as AA_1B and ABA_2, where A_1 and A_2 are points of contact with E_c. From § 2.9, A_1 and A_2 are conjugate points of A. We can obtain a field by restricting ourselves to arcs whose end-points lie between A and its conjugates. For, Fig. VIII. 12, the arc A_1B is then excluded and through B passes only one extremal of the field, namely that arc of the catenary AA_2 which lies between A and A_2.

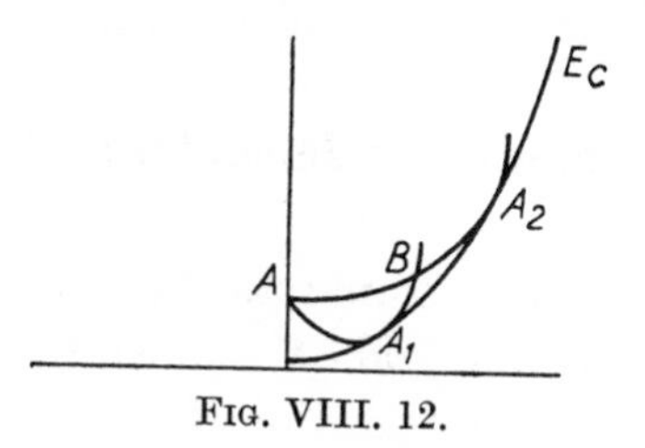

Fig. VIII. 12.

A similar example of a field can be obtained from the integral

$$I = \int (u^2-2gy)^{\frac{1}{2}}(1+y'^2)^{\frac{1}{2}}\,dx. \tag{8}$$

This occurs in the problem of finding the trajectory of a particle by means of the principle of least action, § 1.8. The extremals are a two-parameter family of parabolas

$$c^2(u^2-2gy-c^2) = g^2(x-d)^2,$$

where c and d are arbitrary constants and u is the given initial velocity. We can obtain a one-parameter family by confining

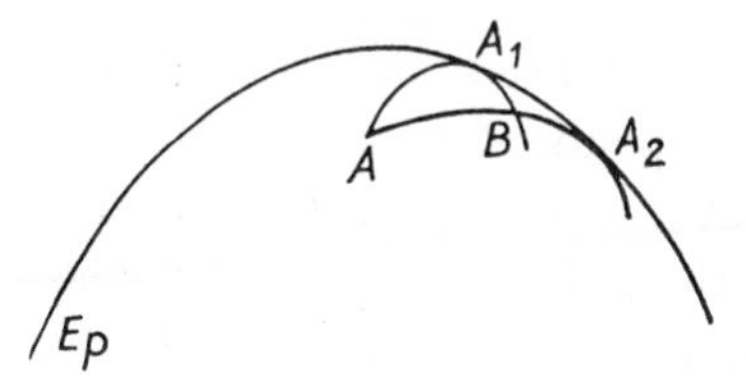

Fig. VIII. 13.

ourselves to those trajectories which lie in a fixed vertical plane through A and which are obtained by projection from A with initial speed u. In § 2.10 it was shown that these trajectories touch a parabola with latus rectum $2u^2/g$ having A as focus and the vertical through A as diameter. If B lies outside the envelope there are no trajectories passing through A and B. If B lies inside there are two trajectories, AA_1B and ABA_2, Fig. VIII. 13, where A_1 and A_2 are the points of contact with the envelope and therefore, § 2.9, are the conjugates of A. As in the previous example, we can obtain a field by confining ourselves to those arcs of the trajectories which lie in between A and its conjugate, for we then have only one extremal through a point B inside the envelope. The field is confined to the interior of a domain bounded by the parabolic envelope and a small curve enclosing A.

8.13. Hilbert's integral independent of the path of integration

In this section we give an analytical proof of the theorem that the value of Hilbert's integral I^*, defined by (3), § 8.12, is independent of the path of integration if we confine ourselves to a field of extremals.

Consider the integral

$$\int_A^B (\alpha\, dx+\beta\, dy), \tag{1}$$

where α and β are functions of x and y. If (i) α, β and their first partial derivatives are continuous and one-valued in a simply connected region of the (x,y) plane, and (ii)

$$\frac{\partial \alpha}{\partial y} = \frac{\partial \beta}{\partial x}, \tag{2}$$

then $\alpha\, dx+\beta\, dy$ is an exact differential and the value of the integral (1) is independent of the path from A to B. These conditions are both necessary and sufficient.†

Consider now the following integral J whose integrand satisfies (i) and (ii) in a field of extremals. J is defined by

$$J = \int_A^B \left\{F+(g'-y')\frac{\partial F}{\partial y'}\right\} dx, \tag{3}$$

where g' is the slope of the path of integration and y' is a function of x and y. Along the path of integration $g'\, dx = dy$, so that we have

$$\alpha = F-y'\frac{\partial F}{\partial y'}; \qquad \beta = \frac{\partial F}{\partial y'}. \tag{4}$$

Hence

$$\frac{\partial \alpha}{\partial y} = \left(\frac{\partial F}{\partial y}+\frac{\partial F}{\partial y'}\frac{\partial y'}{\partial y}\right)-\left(\frac{\partial y'}{\partial y}\frac{\partial F}{\partial y'}+y'\frac{\partial^2 F}{\partial y\, \partial y'}+y'\frac{\partial^2 F}{\partial y'^2}\frac{\partial y'}{\partial y}\right) \tag{5}$$

† R. Courant, *Differential and Integral Calculus*, vol. ii, p. 352.

and (2) becomes

$$\frac{\partial F}{\partial y}-y'\frac{\partial^2 F}{\partial y\partial y'}-y'\frac{\partial^2 F}{\partial y'^2}\frac{\partial y'}{\partial y}=\frac{\partial^2 F}{\partial x\partial y'}+\frac{\partial^2 F}{\partial y'^2}\frac{\partial y'}{\partial x}. \tag{6}$$

Since

$$y''=\frac{dy'}{dx}=\frac{\partial y'}{\partial x}+\frac{\partial y'}{\partial y}y' \tag{7}$$

it is easy to deduce that (6) is equivalent to Euler's characteristic equation

$$\frac{\partial F}{\partial y}-\frac{d}{dx}\left(\frac{\partial F}{\partial y'}\right)=0. \tag{8}$$

The value of J is therefore independent of the path of integration if y' is the slope of an extremal. But this is the case for I^*, the integral defined by (3), § 8.12. Consequently I^* is independent of the path of integration in a field of extremals.

8.14. The method of Carathéodory

A simple method for dealing with the relationship between extremals and transversals has been introduced by Carathéodory.

Consider a one-parameter family of curves and let

$$g(x,y)=a \tag{1}$$

and

$$g(x,y)=a+\delta a \tag{2}$$

be the equations of two neighbouring curves of the family, Fig. VIII. 14. Suppose that we are given (i) a point P on (1), coordinates (x,y), and (ii) a functional form $F(x,y,y')$ where F is known and y' is the derivative of y with respect to x. Consider the problem of finding an arc which minimizes the integral

$$I=\int\limits_P^Q F(x,y,y')\,dx, \tag{3}$$

where the arc starts at P on (1) and terminates at a point Q on (2).

If the arc is sufficiently small I can be expressed as a function of the variable y' as we now proceed to show.

Let the coordinates of Q be $(x+\delta x, y+\delta y)$, then we have

$$I=F(x,y,y')\,\delta x+O(\delta x)^2, \tag{4}$$

where (x,y) are the coordinates of P.

Since Q lies on (2) we have

$$g(x+\delta x, y+\delta y)-g(x,y) = \delta a. \tag{5}$$

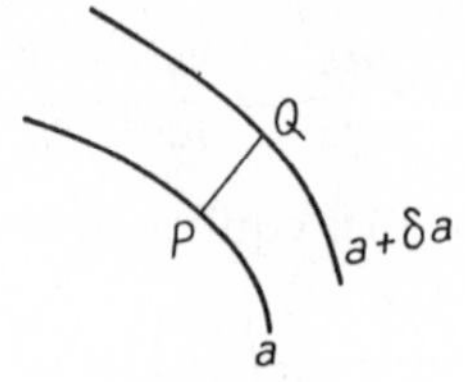

FIG. VIII. 14.

From this we deduce that

$$\left(\frac{\partial g}{\partial x}+y'\frac{\partial g}{\partial y}\right)\delta x = \delta a, \tag{6}$$

where second-order quantities have been ignored. Hence

$$I = \frac{F(x,y,y')}{\dfrac{\partial g}{\partial x}+y'\dfrac{\partial g}{\partial y}}\delta a, \tag{7}$$

where again second-order quantities have been ignored. Evidently y' is the only variable in (7), so that for stationary I we must have $dI/dy' = 0$.

If g' is the slope of (1) at P, i.e. $g' = -\dfrac{\partial g}{\partial x}\Big/\dfrac{\partial g}{\partial y}$, then the condition becomes

$$F+(g'-y')\frac{\partial F}{\partial y'} = 0. \tag{8}$$

Comparison with (10), § 8.2, shows that we have once again arrived at the transversality condition. This may be summed up by the statement that the extremals of (3) must be cut transversally by the family of curves (1).

8.15. The Bliss condition

We can now prove the necessity of condition (vii), Theorem 16, § 8.8. It has been shown in § 8.8 that if the integral I is to be a maximum or minimum and not merely stationary, then focal points must be excluded from the arc of integration AB. The

Bliss condition is additional to this. Let A and B be displaceable along Γ_1 and Γ_2 respectively and let S_1 be the focal point of Γ_1 and S_2 of Γ_2, S_1 and S_2 both lying on the extremal AB. The condition deals with the case when S_1 and S_2 lie on the same side of the arc AB, as in Figs. VIII. 3, VIII. 4, and VIII. 15. If the extremal is traversed in the direction from A to B it can be enunciated as follows:

I can be a maximum or minimum for either of the orders $ABS_2 S_1$ (Fig. VIII. 3) or $S_2 S_1 AB$ (Fig. VIII. 4). I is neither a

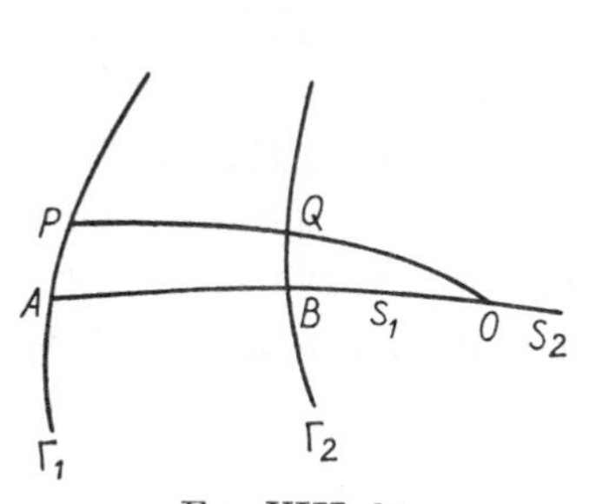

FIG. VIII. 15.

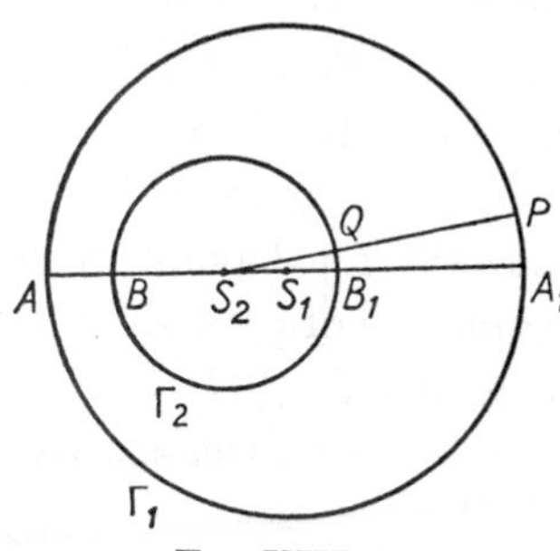

FIG. VIII. 16.

maximum nor a minimum for the orders $ABS_1 S_2$ (Fig. VIII. 15) and $S_1 S_2 AB$.

Before the proof we give an illustration showing the necessity of the condition. Consider the problem of finding the shortest distance between two coplanar circles, Γ_1 and Γ_2. The extremals of the integral to be minimized, $\int ds$, are straight lines and the transversality conditions require these lines to be orthogonal to the circles, example 1, § 8.4. From the results of § 8.9 the focal points of Γ_1 lie on its evolute, which, for a circle, is its centre. Similarly for the focal points of Γ_2. Fig. VIII. 16 shows the two circles Γ_1 and Γ_2 with their centres S_1 and S_2, which are also their respective focal points. If $S_1 S_2$ produced cuts the circles at AB and $A_1 B_1$, as in Fig. VIII. 16, then the common normals AB and $A_1 B_1$ are the stationary positions for $\int ds$. Evidently the Bliss condition is satisfied for AB, so that AB is a minimum distance between the circles. But the condition is not satisfied for $A_1 B_1$, and in fact $A_1 B_1$ is neither a maximum nor a minimum. It is easy to prove this by elementary reasoning.

For on displacing A_1 to P and B_1 to Q so that S_2, P, and Q are collinear, we have $S_2 P < S_2 S_1 + S_1 P = S_2 A_1$ and $S_2 Q = S_2 B_1$. Hence

$$PQ < A_1 B_1.$$

But evidently $A_1 Q > A_1 B_1$ and consequently $A_1 B_1$ can be neither a maximum nor a minimum.

We now proceed with the proof of the Bliss condition. We assume that all the conditions for minimum I are satisfied except that the focal points are in the order exhibited in Fig. VIII. 15, and we then show that this leads to a contradiction. For brevity $\int F(x, y, y')\,dx$ taken along the arc ABC will be denoted by $I(AC)$ or by $I(ABC)$.

Take any point O between S_1 and S_2 in Fig. VIII. 15. Then, since the focal point S_1 is not excluded from the arc AO, the arguments of § 8.8 show that we can find a curve such as PQO for which $I(PQO) < I(ABO)$. But since the focal point S_2 is excluded from the arc BO, we have

$$I(QO) > I(BO).$$

Consequently $I(PQ) < I(AB)$, which contradicts the original assumptions. Thus the necessity for the Bliss condition is proved.

CHAPTER IX

STRONG VARIATIONS AND THE WEIERSTRASSIAN E FUNCTION

9.1. Introduction

WEAK variations have already been defined in § 1.3. If $y = s(x)$ is the equation of an extremal Γ_e and $y = s(x)+\epsilon t(x)$ that of a neighbouring curve Γ_n then, in the case of weak variations, $t(x)$ is independent of ϵ and so $\epsilon t'(x)$ tends to zero with ϵ. Alternatively if, as ϵ tends to zero, a point Q on Γ_n approaches P on Γ_e, then the slope of Γ_n at Q must tend to that of Γ_e at P. This excludes an important type of variation, as the following example shows.

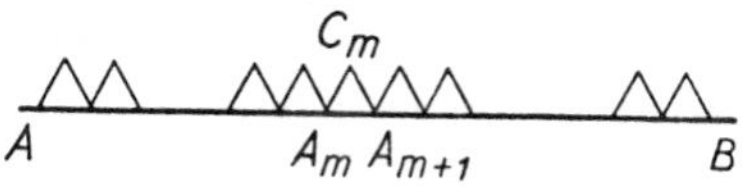

FIG. IX. 1.

Consider the problem of finding the shortest distance between two points A and B, for which the extremal Γ_e is the straight line AB (Fig. IX. 1). Divide AB into n equal intervals, each of length AB/n, and let $A_m A_{m+1}$ ($A_1 = A$, $A_{n+1} = B$) denote the mth interval. Construct the equilateral triangles $A_m A_{m+1} C_m$ ($m = 1, 2,..., n$) and for the neighbouring curve Γ_n take the series of lines $A_m C_m$ and $C_m A_{m+1}$ ($m = 1, 2,..., n$). Evidently as n tends to infinity each point on Γ_n tends to some point on Γ_e but the slope of Γ_n always differs from that of Γ_e by $\pi/3$. The length of Γ_n is twice that of Γ_e, so that the straight line AB still gives us the minimum case.

The limitation imposed upon variational theory by the use of weak variations was first transcended by Weierstrass who, in the year 1879, introduced an expression now fundamental in variational theory. The study of this expression, generally known as the Weierstrassian E function, is the main purpose of the present chapter. The E function enables us to deal with the most general possible variations from Γ_e to Γ_n.

9.2. The Weierstrassian E function in the simplest case

Suppose that with the integral

$$I = \int_A^B F(x,y,y')\,dx = \int_A^B F\,dx, \tag{1}$$

whose end points A and B are fixed, we can associate a field of extremals, as defined in § 8.12. Then, § 8.13, the value of the Hilbert integral

$$I^* = \int_A^B \left\{F+(g'-y')\frac{\partial F}{\partial y'}\right\} dx \tag{2}$$

is the same for all paths from A to B confined to the domain of the field. If the path is along the extremal Γ_e, then $g' = y'$ so that

$$I^* = I. \tag{3}$$

Let P and Q be two points, one on Γ_e and the second on a neighbouring curve Γ_n, Fig. IX. 2, and let their respective coordinates be (x,y) and (x,Y). The slope of Γ_n at Q will be denoted by Y'. Taken along the path Γ_n we have

$$I^* = \int_A^B \left\{F(x,Y,y')+(Y'-y')\frac{\partial F(x,Y,y')}{\partial y'}\right\} dx, \tag{4}$$

where y' is the slope of the extremal through Q.

From the invariant properties of I^* it follows that

$$\int_A^B F(x,y,y')\,dx = \int_A^B \left\{F(x,Y,y')+(Y'-y')\frac{\partial F(x,Y,y')}{\partial y'}\right\} dx, \tag{5}$$

where the left-hand integral is taken along Γ_e and the right-hand one along Γ_n.

In (1) we take I to be the value of the integral when the path from A to B is along the extremal Γ_e, and $I+\delta I$ to be its value when taken along the neighbouring path Γ_n. Then

$$I+\delta I = \int_A^B F(x,Y,Y')\,dx, \tag{6}$$

and so from (5) we have

$$\delta I = \int_A^B \left\{F(x,Y,Y')-F(x,Y,y')-(Y'-y')\frac{\partial F(x,Y,y')}{\partial y'}\right\} dx, \tag{7}$$

where Γ_n is the path of integration.

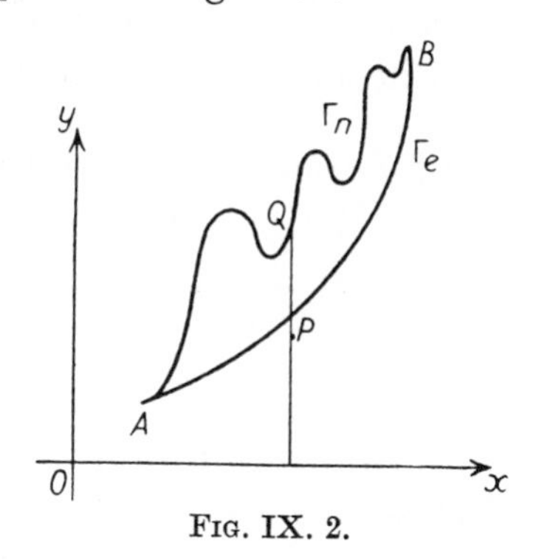

FIG. IX. 2.

The integrand of (7) is the Weierstrassian E function for the integral (1), and on writing

$$E(x,y,y',Y') = F(x,y,Y')-F(x,y,y')-(Y'-y')\frac{\partial F(x,y,y')}{\partial y'} \tag{8}$$

(7) becomes

$$\delta I = \int_A^B E(x,Y,y',Y')\,dx. \tag{9}$$

In this integral the path is taken along Γ_n, a curve which lies in the field of extremals containing Γ_e. The coordinates of a point Q on Γ_n are (x, Y), the slope of Γ_n at Q is Y', and that of the extremal passing through Q is y'.

As Q on Γ_n tends to P on Γ_e, Y tends to y but Y' need not necessarily tend to y'. A simple case is given by the example of § 9.1. The variations of this section are therefore much more general than those of § 1.3 and are accordingly known as *strong variations*. Associated minima (maxima) are known as strong minima (maxima). It is evident from (9) that a necessary condition for a strong minimum is that

$$E(x,y,y',Y') \geqslant 0 \tag{10}$$

at all points of the extremal and for all finite values of Y' (we exclude the possibility that $E(x,y,y',Y') = 0$ at all points of the extremal). For a strong maximum the inequality sign must be reversed. Inequality (10) is known as the Weierstrassian condition.

In addition to this there are, of course, other conditions such as the Eulerian equation and the exclusion of conjugate points; these will appear later in the general discussion of § 9.6.

9.3. The simplified form of the Weierstrassian condition

If E has continuous partial derivatives of the first and second order we may apply the second mean value theorem† to (8), § 9.2. We then have

$$E(x,y,y',Y') = \tfrac{1}{2}(Y'-y')^2 \frac{\partial^2 F\{x,y,y'+\theta(Y'-y')\}}{\partial y'^2}, \tag{1}$$

where $0 < \theta < 1$. Thus the inequality (10), § 9.2, is ensured if

$$\frac{\partial^2 F(x,y,p)}{\partial p^2} \geqslant 0 \tag{2}$$

at every point of the extremal Γ_e and for all finite values of p. It is possible, however, for (10), § 9.2, and (1) to be true for all values of Y', but for (2) to be true for only a limited range of values of p. Thus (2), although useful in practice and sufficient to ensure the truth of (10), § 9.2, is not a necessary condition.

The inequality (2) must be carefully distinguished from the Legendre test of § 1.5 and § 2.5, which requires that $\partial^2 F/\partial y'^2$ should have a constant sign at all points of the extremal Γ_e. In the Legendre test y' is the slope of Γ_e, whereas in (2) the variable p can assume any finite value.

To ensure a maximum value for I the inequality sign of (2) must be reversed.

EXAMPLE 1. Investigate the case when

$$I = \int_A^B f(x,y)\,ds = \int_A^B f(x,y)(1+y'^2)^{\frac{1}{2}}\,dx, \tag{3}$$

where the positive value of the root only is taken.

† R. Courant, *Differential and Integral Calculus*, vol. ii, p. 80.

Here
$$\frac{\partial^2 F(x,y,p)}{\partial p^2} = \frac{f(x,y)}{(1+p^2)^{\frac{3}{2}}}, \tag{4}$$
so that I possesses a strong minimum if $f(x,y) \geqslant 0$ at all points of the extremal Γ_e (excluding the case when $f(x,y) = 0$ all along Γ_e).

This result includes the following special cases:

(i) The shortest distance between two points, where the integral to be minimized is $\int\limits_A^B ds$. Here $f(x,y) = 1$ and the integral therefore admits a strong minimum.

(ii) Problems associated with catenaries, § 1.7, and minimal surfaces, § 1.12. The relevant integral is $\int\limits_A^B y\, ds$, so that the condition for a strong minimum is satisfied if $y > 0$ and the arc of the catenary remains wholly above the x-axis.

(iii) The principle of least action for a particle in a conservative field of force, § 1.8. The integral is $\int\limits_A^B v\, ds$, where v is the speed. If $v > 0$ throughout the path, the condition for a strong minimum is realized.

(iv) Problems associated with optics, § 1.9, and the brachistochrone, § 1.11. The integral is $\int\limits_A^B ds/v$, where v is the speed. A strong minimum is realized once again if $v > 0$ at all points of the optical or dynamical path.

(v) Geodesics satisfy the requirements for strong minima. The general case is more easily dealt with by expressing the integral in parametric form, as in § 9.14, and in this section we shall confine ourselves to the case of geodesics on a sphere. The relevant integral, equation (2), § 1.10, is
$$I = \int\limits_A^B (1+y'^2 \sin^2 x)^{\frac{1}{2}}\, dx, \tag{5}$$
where the positive value of the root only is taken. We have
$$\frac{\partial^2 F}{\partial p^2} = \frac{\sin^2 x}{(1+p^2 \sin^2 x)^{\frac{3}{2}}}, \tag{6}$$

so that $\partial^2F/\partial p^2 > 0$ except at $x = 0$ and $x = \pi$ where it vanishes.

EXAMPLE 2. Newton's solid of minimum resistance. This problem was discussed in § 1.16 and the integral to be minimized is

$$I = \int \frac{x}{1+y'^2}\,dx. \tag{7}$$

For this case we have

$$\frac{\partial^2 F}{\partial p^2} = -\frac{2x(1-3p^2)}{(1+p^2)^3}, \tag{8}$$

so that $\partial^2F/\partial p^2$ cannot maintain a constant sign for all values of p. Since this test is sufficient but not necessary, we return to the necessary E test, (10), § 9.2. From (8), § 9.2, we have

$$E(x,y,y',Y') = \frac{x(Y'-y')^2(2Y'y'+y'^2-1)}{(1+Y'^2)(1+y'^2)^2}. \tag{9}$$

Evidently the inequality $E(x,y,y',Y') \geqslant 0$ is not true for all finite values of Y', so that the integral possesses a weak but not a strong minimum.

The extremals of (7) are given by eliminating the parameter p from equations (4) and (6), § 1.16, and the solid of minimum resistance, S_m, is then obtained by revolving these curves about the y-axis. The investigation of this section shows that it must be possible to construct solids of revolution, S_{mm}, which offer less resistance to fluids than S_m. The meridian curves of S_{mm} would not possess continuously varying tangents, as is the case for S_m, but would be built up from a series of arcs which change direction discontinuously at their points of intersection, as in the discontinuous solutions of § 1.17.

9.4. The Weierstrassian condition by an alternative method

The Eulerian equation of the integral

$$I = \int\limits_A^B F(x,y,y')\,dx = \int\limits_A^B F\,dx \tag{1}$$

defines a family of extremals with two parameters. Let Γ_e denote the extremal which passes through A and B, the fixed end points of the arc of integration, and let us assume that there are no points conjugate to A or B lying within the arc. Take P and Q, any two points on the arc between A and B, and through P draw any curve Γ making an angle other than zero with Γ_e at P (Fig. IX. 3). Through Q can be drawn a singly infinite one-parameter family of extremals of (1). Let one of these meet Γ at R.

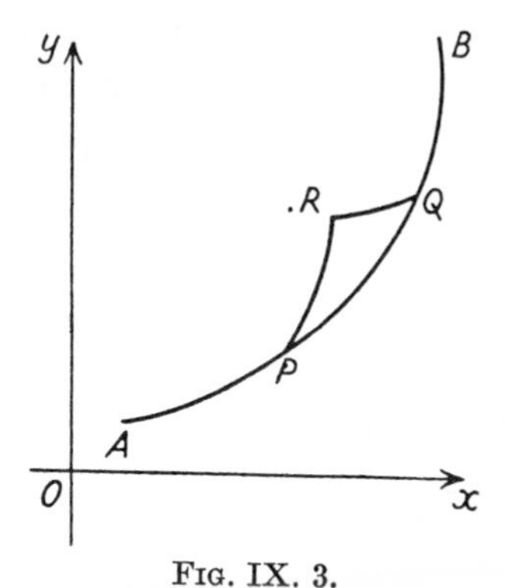

FIG. IX. 3.

Consider the integral

$$J = \int_A^P F\,dx + \int_P^R F\,dx + \int_R^Q F\,dx + \int_Q^B F\,dx. \tag{2}$$

As R moves along Γ, J will be a function of ξ, the abscissa of R, and we may denote it by $J(\xi)$. If the abscissa of P is x_1 it is evident that $J(x_1) = I$.

If I is to be a minimum for the path Γ_e, then J must be a non-decreasing function of ξ as R moves away from P, so that $\partial J/\partial\xi \geqslant 0$ when $\xi = x_1$. In interpreting this result we must look upon $\delta\xi$ as the radius vector of R in a system of polar co-ordinates of which P is the pole and the tangent to AB at P is the initial line. The direction from A to B along the extremal will be called the forward direction. In Fig. IX. 3, as Q tends to P the direction of PQ tends to that of the forward tangent at P. The vectorial angle of PR will be the angle measured counter-clockwise from the forward tangent at P.

We now proceed to evaluate $(\partial J/\partial\xi)_p$. Evidently the derivatives of the first and fourth integrals of (2) vanish for fixed P and

Q. The derivative of the second integral is $F(x, y, Y')$, where Y' is the slope of PR at P. Since QR is an extremal with variable end point R the derivative of the third integral can be obtained by the method of § 8.2. From (9), § 8.2, the result is

$$-\left\{F+(Y'-y')\frac{\partial F}{\partial y'}\right\}_p, \tag{3}$$

where (i) the suffix p denotes that all variables have their values taken at the point P, (ii) y' is the slope of the forward tangent to the extremal at P, and (iii) Y' is the slope of Γ at P.

Collecting these results we have

$$\left(\frac{\partial J}{\partial \xi}\right)_p = \left\{F(x,y,Y')-F(x,y,y')-(Y'-y')\frac{\partial F(x,y,y')}{\partial y'}\right\}_p \tag{4}$$

$$= E(x,y,y',Y')_p, \tag{5}$$

from (8), § 9.2. Consequently Γ_e is the path for minimum I if $E(x,y,y',Y') \geqslant 0$ at all points of Γ_e and for all finite values of Y', in agreement with (10), § 9.2.

9.5. Conjugate points related to fields of extremals

The object of this section is to show that there is a close connexion between conjugate points, defined in § 2.6, and fields of extremals, defined in § 8.12. The results obtained so far can then be expressed in terms of whichever of these two concepts is more convenient. We shall prove the following result.

Let AB be an arc of an extremal for the integral (1) of § 9.4. If AB can be enclosed in an arc $A_0 B_0$ which contains no points conjugate to A and B and no zeros of $\partial^2 F/\partial y'^2$, then we can construct a field of extremals, as defined in § 8.12, of which the extremal AB is a member.

To prove this we first observe that the Eulerian equation for the integral (1), § 9.4, has solutions of the form

$$y = s(x, c_1, c_2), \tag{1}$$

where c_1 and c_2 are parameters. Associated with this is the accessory equation, (1), § 2.4, a second-order equation having $\partial y/\partial c_1$ and $\partial y/\partial c_2$ as independent solutions, § 2.8. Let P be a point on the extremal Γ_e, then the solution of the accessory equation which vanishes at P also vanishes at all points on Γ_e which are conjugate to P, § 2.6.

Let (x_0, y_0) be the coordinates of a point P on Γ_e. Then the one-parameter family of extremals passing through P can be obtained by eliminating one of the parameters c_1 or c_2 from (1) and

$$y_0 = s(x_0, c_1, c_2). \tag{2}$$

A more symmetrical result is obtained if c_1 and c_2 are expressed in terms of a third parameter α, defined by

$$\alpha = \frac{\partial s(x_0, c_1, c_2)}{\partial x_0}. \tag{3}$$

The one-parameter family of extremals through P then takes the form

$$y = \phi(x, \alpha). \tag{4}$$

From (1) we have

$$\frac{\partial y}{\partial \alpha} = \frac{\partial y}{\partial c_1}\frac{\partial c_1}{\partial \alpha} + \frac{\partial y}{\partial c_2}\frac{\partial c_2}{\partial \alpha}. \tag{5}$$

Differentiating (2) and (3) with respect to α we can then eliminate $\partial c_1/\partial\alpha$ and $\partial c_2/\partial\alpha$ from (5). The result is

$$\frac{\partial y}{\partial \alpha} = C\left\{-\frac{\partial y}{\partial c_1}\frac{\partial y_0}{\partial c_2} + \frac{\partial y}{\partial c_2}\frac{\partial y_0}{\partial c_1}\right\}, \tag{6}$$

where C is a constant depending upon x_0 and y_0. On writing $u_1 = \partial y_0/\partial c_1$, $u_1' = du_1/dx_0$, $u_2 = \partial y_0/\partial c_2$ and $u_2' = du_2/dx_0$ we have

$$\frac{1}{C} = u_1 u_2' - u_1' u_2. \tag{7}$$

Since $\partial y/\partial c_1$ and $\partial y/\partial c_2$ are solutions of the accessory equation it follows from (4), § 2.18, that (7) can be written in the form

$$C = K\frac{\partial^2 F}{\partial y'^2}, \tag{8}$$

where K is a constant independent of x_0 and y_0. But, by hypothesis, $\partial^2 F/\partial y'^2$ does not vanish at any point of an arc $A_1 B_1$, enclosed by $A_0 B_0$ and enclosing AB, see Fig. IX. 4. Therefore C cannot vanish at any point of $A_1 B_1$.

Now, from (6), $(1/C)\partial y/\partial\alpha$ is a solution of the accessory equation which vanishes at P, where $y = y_0$, and consequently its zeros can occur only at the conjugates of P. Hence, if P_c is the nearest conjugate, $\partial y/\partial\alpha$ has constant sign for that section of PP_c which lies inside $A_1 B_1$ (Fig. IX. 4). By choosing P sufficiently near to A, but outside the arc AB, and observing that the conjugate of A lies outside AB, we deduce that $\partial y/\partial\alpha$ has constant

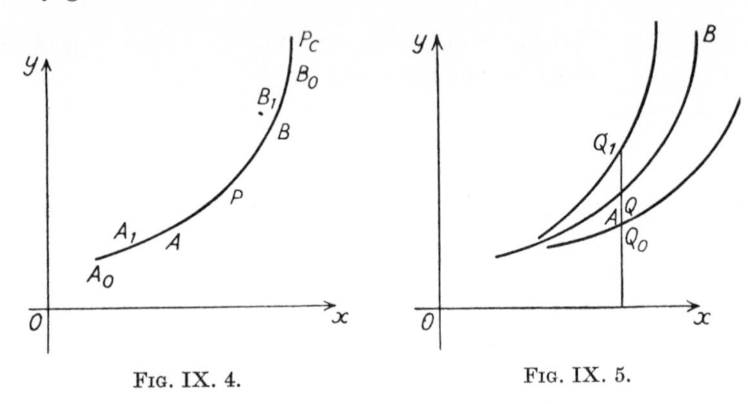

FIG. IX. 4. FIG. IX. 5.

sign and cannot vanish for some arc $A_2 B_2$ which encloses AB. Assuming that $\partial y/\partial\alpha$ is a continuous function of x and α, this constancy of sign must hold at all points of a strip or domain S which encloses AB. Thus, if a member of (4) possesses an arc CD which lies in the domain S, $\partial y/\partial\alpha$ has constant sign at all points of CD. In constructing S we must exclude the point P.

We can now prove that those arcs of (4) which lie within S possess the properties of a field of extremals. For let Q be a point in S (Fig. IX. 5) and let the ordinate through Q intersect the boundaries of S at Q_0 and Q_1. As Q, coordinates (x_1, y_1), traverses the interval $Q_0 Q_1$, x_1 remains constant and (4) becomes a relation between y_1 and α, namely $y_1 = \phi(x_1, \alpha)$. But the constancy of sign of $\partial y/\partial\alpha$ shows that as y steadily increases α either steadily increases or steadily decreases. Hence, for given y_1, the equation for α, $\phi(x_1, \alpha) = y_1$, can have only one root, and only one member of (4) can therefore pass through Q. Consequently the domain S encloses a field of extremals of which the extremal AB is a member, as required.

9.6. Conditions for a strong maximum or minimum

The results so far obtained can be summarized in the form of two theorems. The integral I, with fixed end points A and B for the range of integration, is defined as usual by

$$I = \int_A^B F(x,y,y')\,dx = \int_A^B F\,dx. \tag{1}$$

THEOREM 17. *The following conditions are necessary to ensure a strong minimum for the integral I.*

(i) *The equation of Γ_e, the arc of integration, must satisfy*

$$\frac{\partial F}{\partial y} - \frac{d}{dx}\left(\frac{\partial F}{\partial y'}\right) = 0.$$

(ii) *The arc AB, of Γ_e, contains no point conjugate to either A or B* (see § 2.6).

(iii) *At all points of this arc* $\dfrac{\partial^2 F}{\partial y'^2} > 0.$

(iv) *At all points of this arc and for all finite values of p*

$$E(x,y,y',p) \geqslant 0$$

(see § 9.2). *The possibility $E(x,y,y',p) = 0$ at all points of the arc for all finite values of p is excluded.*

For a strong maximum the inequality signs of (iii) *and* (iv) *must be reversed.*

THEOREM 18. *The following conditions are sufficient to ensure a strong minimum for the integral I.*

(i) *and* (ii) *as for Theorem* 17.

(iii *a*) *At all points of the arc AB, of Γ_e, and for all finite values of p*

$$\frac{\partial^2 F(x,y,p)}{\partial p^2} > 0.$$

For a strong maximum the inequality sign of (iii *a*) *must be reversed.*

The difference between conditions (iii) and (iii *a*) has already been noted. In Theorem 18 (iii *a*) p can have any value, whereas in Theorem 17 (iii) y' is the slope of the extremal.

Conditions (ii) and (iii), or (ii) and (iii *a*), enable us to construct a field of extremals containing Γ_e as a member. The proof of Theorem 17 then follows immediately from (iv) and the analysis of § 9.2, in particular (9), § 9.2. The proof of Theorem 18 follows from (1), § 9.3, for if (iii *a*) holds, then conditions (iii) and (iv) of Theorem 17 are both satisfied.

But it is possible for (iii) and (iv) to be satisfied without (iii *a*) holding for *all* values of p, so that the conditions of Theorem 18 are sufficient but not necessary.

Evidently if I has a strong minimum it must also have a weak one, but the converse is not necessarily true, as the problem of the solid of minimum resistance, § 9.3, shows. Weak variations are very special cases of strong ones, so that a weak minimum can hardly be regarded as a true minimum. Nevertheless the possession of a weak minimum is an important functional property which may have useful applications to practical problems. Much of the analysis which occurs in the theory of weak variations still retains its value in the theory of strong variations, e.g. accessory equations, conjugate points, focal points, etc. It is mainly the Legendre test of §§ 1.5 and 2.5 which is seriously affected by the Weierstrassian introduction of strong variations, for this must be replaced by the far more powerful E test of § 9.2.

9.7. Strong variations for integrals with two dependent variables

The results obtained above for integrals with one dependent variable can be generalized in many ways. In this section we shall state, without proof, the generalization to integrals of the type

$$I = \int\limits_A^B H\left(x, y, z; \frac{\partial y}{\partial x}, \frac{\partial z}{\partial x}\right) dx = \int\limits_A^B H\,dx, \tag{1}$$

where the path of integration is along a twisted curve in three-dimensional space. Here x is the independent and y, z are the dependent variables. We confine ourselves to the case where the end points of integration, A and B, are fixed, and for brevity we write $p = \partial y/\partial x$ and $q = \partial z/\partial x$.

I is stationary if y and z, as functions of x, satisfy the two Eulerian equations†

$$\left.\begin{aligned}\frac{\partial H}{\partial y}-\frac{d}{dx}\left(\frac{\partial H}{\partial p}\right)&=0,\\ \frac{\partial H}{\partial z}-\frac{d}{dx}\left(\frac{\partial H}{\partial q}\right)&=0.\end{aligned}\right\} \tag{2}$$

Theories of conjugate points and fields of extremals can be developed for (1) in much the same way as for the integral $\int F\left(x,y,\frac{dy}{dx}\right)dx$. The corresponding Hilbert integral I^* is given by

$$I^* = \int\limits_A^B \left\{H\,dx+(dy-p\,dx)\frac{\partial H}{\partial p}+(dz-q\,dx)\frac{\partial H}{\partial q}\right\}dx. \tag{3}$$

At every point X of the path there is a tangent T_p to the path and a tangent T_e to the extremal of the field through X. The direction ratios of T_p and T_e are respectively $(dx\!:\!dy\!:\!dz)$ and $(1\!:\!p\!:\!q)$. In a field of extremals it can be proved that the value of I^* is independent of the path of integration and depends upon the coordinates of the end points only.‡

If at a point P of a field of extremals we have

$$H\,dx+(dy-p\,dx)\frac{\partial H}{\partial p}+(dz-q\,dx)\frac{\partial H}{\partial q}=0, \tag{4}$$

then the direction $(dx\!:\!dy\!:\!dz)$ is said to be transversal to the extremal through P. All transversal directions at P evidently lie in a plane and the envelopes of these planes are surfaces known as transversals of the field.

The Weierstrassian E function for (1) is given by

$$E(x,y,z;\,p,q;\,P,Q) = H(x,y,z;\,P,Q)-H(x,y,z;\,p,q)-$$
$$-(P-p)\frac{\partial H(x,y,z;\,p,q)}{\partial p}-(Q-q)\frac{\partial H(x,y,z;\,p,q)}{\partial q}. \tag{5}$$

Let Γ_e denote the extremal through A and B and Γ_n a neighbouring curve passing through these points. If δI denotes the

† See § 3.2, equations (7) and (8).

‡ For proofs of these and subsequent statements see G. A. Bliss, *Lectures on the Calculus of Variations*, University of Chicago Press. For conjugate points and fields, pp. 29 and 46; for Hilbert's integral, pp. 44 and 49; for extensions to space of $(n+1)$ dimensions, p. 86.

difference between the values of I when the paths of integration are Γ_n and Γ_e, it can be proved that

$$\delta I = \int_A^B E(x, y, z; p, q; P, Q)\, dx, \tag{6}$$

where the integration is along Γ_n. At a point X of Γ_n the direction ratios $(1:P:Q)$ and $(1:p:q)$ are respectively those of the tangent to Γ_n and the tangent to the extremal of the field passing through X.

In a field of extremals we can obtain generalizations of theorems 17 and 18. Thus I has a strong minimum if (i) equations (2) are satisfied, (ii) the arc AB of Γ_e contains no point conjugate to A or B, (iii) $E(x, y, z; p, q; P, Q) \geqslant 0$ at every point of this arc and for every finite value of P and Q.†

9.8. The Weierstrassian theory for integrals in parametric form

Curves can often be studied with great facility when the coordinates of their points are expressed in parametric form. For example the coordinates of points on the circle $x^2+y^2 = a^2$ can be expressed in the form $(a\cos t, a\sin t)$ or alternatively in the form $\left(a\frac{1-\tau^2}{1+\tau^2}, a\frac{2\tau}{1+\tau^2}\right)$. An infinite number of such parametric representations are possible for any curve.

In the general case we shall assume that the coordinates (x, y) can be expressed in parametric form by means of the equations

$$x = \phi(t), \qquad y = \psi(t), \tag{1}$$

and we shall denote differentiation with respect to t by means of dots.‡ The integral

$$I = \int_A^B F\left(x, y, \frac{dy}{dx}\right) dx \tag{2}$$

then becomes

$$I = \int_A^B F\left(x, y, \frac{\dot{y}}{\dot{x}}\right)\dot{x}\, dt. \tag{3}$$

† Bliss, loc. cit. For proof of (6) see p. 41 and for conditions for a strong minimum see p. 43.

‡ No association with time is implied by the use of t.

We shall write this in the form

$$I = \int_A^B G(x, y, \dot{x}, \dot{y})\, dt \tag{4}$$

and observe that the integrand of (4) is a homogeneous function of degree one in the variables $\dot{x}$, $\dot{y}$.

Evidently the value of I must be independent of the parametric form chosen for x and y. For example the length of the circular arc,

$$\int ds = \int (\dot{x}^2+\dot{y}^2)^{\frac{1}{2}}\, dt, \tag{5}$$

must be the same whether we use $x = a\cos t$, $y = a\sin t$ or $x = a\left(\frac{1-t^2}{1+t^2}\right)$, $y = a\left(\frac{2t}{1+t^2}\right)$, as is easily verified. If we write $t = 2\tan^{-1}\tau$ in the first pair of these equations we obtain the second pair with t replaced by τ. Our first problem then is to determine the conditions which ensure that (2) remains invariant whatever parametric form is chosen.

The change from one parameter to another may be expressed by a relationship of the form

$$t = \phi(\tau), \tag{6}$$

where τ is the parameter after the change. Thus we require the conditions under which (4) remains invariant for the transformation (6). For such invariance we must have

$$\int_{t_1}^{t} G\left(x, y, \frac{dx}{dt}, \frac{dy}{dt}\right) dt = \int_{\tau_1}^{\tau} G\left(x, y, \frac{dx}{d\tau}, \frac{dy}{d\tau}\right) d\tau, \tag{7}$$

where t_1 and τ_1 correspond to the point A and t and τ correspond to any point on the arc of integration.

On differentiating (7) we have

$$G\left(x, y, \frac{dx}{dt}, \frac{dy}{dt}\right)\frac{dt}{d\tau} = G\left(x, y, \frac{dx}{d\tau}, \frac{dy}{d\tau}\right). \tag{8}$$

In the case of the special transformation

$$t = k\tau \quad (k \text{ positive and constant}) \tag{9}$$

this becomes

$$G\left(x, y, \frac{dx}{dt}, \frac{dy}{dt}\right)k = G\left(x, y, k\frac{dx}{dt}, k\frac{dy}{dt}\right). \tag{10}$$

The invariance of (4) under transformation (6) must hold not merely along a given arc but in some domain D enclosing the arc AB. In such a domain we can choose paths of integration for which x, y, $\dot{x}$, $\dot{y}$ can assume arbitrary values. If k is other than unity, (10) can hold only if $G(x,y,\dot{x},\dot{y})$ is homogeneous and of degree one in $\dot{x}$ and $\dot{y}$.

Conversely if $G(x,y,\dot{x},\dot{y})$ is homogeneous and of degree one in the variables $\dot{x}$ and $\dot{y}$, equation (8) is obviously true. If $dt/d\tau > 0$ we may then integrate with respect to τ and deduce (7).

Consequently the homogeneity condition is necessary and sufficient to ensure the invariance of I whatever parametric form is chosen. For the rest of this chapter we confine ourselves entirely to functions satisfying this condition.

9.9. The Eulerian equation for $\int_{t_1}^{t_2} G(x, y, \dot{x}, \dot{y})\, dt$

The great advantage of the parametric form is that x and y can be varied independently of each other. Let P, coordinates (x,y), be a point on Γ_e, the path of integration which renders the integral

$$I = \int_{t_1}^{t_2} G(x,y,\dot{x},\dot{y})\, dt = \int_{t_1}^{t_2} G\, dt \tag{1}$$

stationary. Consider a path of integration along a neighbouring curve Γ_n obtained by displacing P to the point Q, coordinates $x+\epsilon_1 \xi_1(t)$, $y+\epsilon_2 \xi_2(t)$. Here ϵ_1 and ϵ_2 are constants and $\xi_1(t)$, $\xi_2(t)$ are arbitrary functions of t independent of ϵ_1 and ϵ_2. Restricting ourselves to the case of fixed end points we have

$$\xi_1(t_1) = \xi_2(t_1) = \xi_1(t_2) = \xi_2(t_2) = 0.$$

The arguments used in § 3.2 can be repeated here and we then prove that I is stationary if x and y, regarded as functions of t, satisfy the equations

$$\frac{\partial G}{\partial x} - \frac{d}{dt}\left(\frac{\partial G}{\partial \dot{x}}\right) = 0, \tag{2}$$

$$\frac{\partial G}{\partial y} - \frac{d}{dt}\left(\frac{\partial G}{\partial \dot{y}}\right) = 0. \tag{3}$$

When G satisfies the homogeneity conditions these equations are not independent of each other, as we now proceed to prove.

Since G is homogeneous and of degree one in $\dot{x}$ and $\dot{y}$ we have

$$\dot{x}\frac{\partial G}{\partial \dot{x}}+\dot{y}\frac{\partial G}{\partial \dot{y}} = G. \tag{4}$$

Differentiating this identity first with respect to $\dot{x}$ and then to $\dot{y}$ proves that

$$\frac{1}{\dot{y}^2}\frac{\partial^2 G}{\partial \dot{x}^2} = -\frac{1}{\dot{x}\dot{y}}\frac{\partial^2 G}{\partial \dot{x}\partial \dot{y}} = \frac{1}{\dot{x}^2}\frac{\partial^2 G}{\partial \dot{y}^2} = S(x,y,\dot{x},\dot{y}) = S, \tag{5}$$

where $S(x,y,\dot{x},\dot{y})$ denotes the common value of each of these expressions. Again on differentiating (4) partially with respect to x we obtain

$$\frac{\partial G}{\partial x} = \dot{x}\frac{\partial^2 G}{\partial x\partial \dot{x}}+\dot{y}\frac{\partial^2 G}{\partial x\partial \dot{y}}. \tag{6}$$

Now apply (5) and (6) to (2). It follows that

$$\frac{\partial G}{\partial x}-\frac{d}{dt}\left(\frac{\partial G}{\partial \dot{x}}\right) = \frac{\partial G}{\partial x}-\frac{\partial^2 G}{\partial x\partial \dot{x}}\dot{x}-\frac{\partial^2 G}{\partial \dot{x}^2}\ddot{x}-\frac{\partial^2 G}{\partial y\partial \dot{x}}\dot{y}-\frac{\partial^2 G}{\partial \dot{y}\partial \dot{x}}\ddot{y}$$

$$= \dot{y}\left\{\frac{\partial^2 G}{\partial x\partial \dot{y}}-\frac{\partial^2 G}{\partial y\partial \dot{x}}-S(\dot{y}\ddot{x}-\dot{x}\ddot{y})\right\}. \tag{7}$$

Similarly we can prove that

$$\frac{\partial G}{\partial y}-\frac{d}{dt}\left(\frac{\partial G}{\partial \dot{y}}\right) = -\dot{x}\left\{\frac{\partial^2 G}{\partial x\partial \dot{y}}-\frac{\partial^2 G}{\partial y\partial \dot{x}}-S(\dot{y}\ddot{x}-\dot{x}\ddot{y})\right\}. \tag{8}$$

Since $\dot{x}$ and $\dot{y}$ do not in general vanish together† the two equations (2) and (3) are equivalent to the single equation

$$\frac{\partial^2 G}{\partial x\partial \dot{y}}-\frac{\partial^2 G}{\partial y\partial \dot{x}}-S(x,y,\dot{x},\dot{y})(\ddot{x}\dot{y}-\ddot{y}\dot{x}) = 0. \tag{9}$$

This is known as the Weierstrassian form of the Eulerian equation. To use (9), first express x in terms of any convenient function of t, substitute in (9), and solve for y in terms of t. Alternatively choose any relationship between x, y, and t and by means of (9) x and y can then be expressed in terms of t.

In the subsequent analysis we shall sometimes take t to be the length of arc and $(\dot{x},\dot{y})$ are then the direction cosines of the tangent to the extremal at the point (x,y).

† $\dot{x}$ and $\dot{y}$ vanish simultaneously at singular points. We shall assume that the extremal is free from such points when t lies inside the interval (t_1, t_2).

9.10. The Weierstrassian E function for $\int_A^B G(x, y, \dot{x}, \dot{y})\,dt$

Let Γ_e be the extremal and Γ_n a neighbouring curve. Our present aims are firstly to obtain δI, the difference between the values of $\int G\,dt$ taken along Γ_n and Γ_e, and secondly to express the result in a form analogous to (7), § 9.2. For these purposes we take a convenient point O on Γ_e along the arc BA produced, Fig. IX. 6, and draw the family of extremals which pass through O. Let two neighbouring members of this family intersect Γ_n at Q_m and Q_{m+1}, Fig. IX. 6, and let $I(MN)$ denote the value of $\int G(x,y,\dot{x},\dot{y})\,dt$ when integrated along the arc MN. Then it is easy to see from the figure that

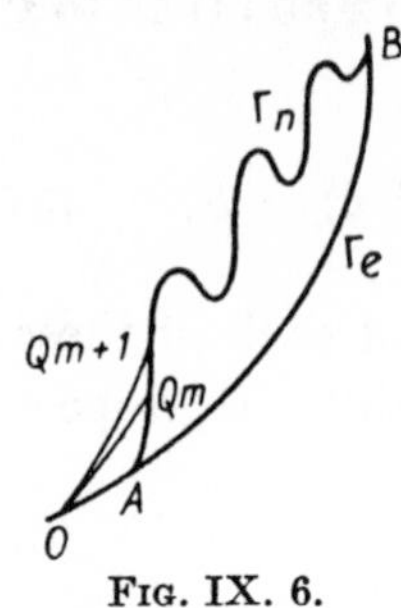

FIG. IX. 6.

$$\delta I = \sum_{m=1}^{n} \{I(OQ_m)+I(Q_m\,Q_{m+1})-I(OQ_{m+1})\}, \tag{1}$$

where $Q_1 = A$ and $Q_{n+1} = B$.

The evaluation of the right-hand side of (1) can be much shortened by using the results of § 8.2, where a difference similar to $I(OQ_{m+1})-I(OQ_m)$ has already been obtained. We use (9), § 8.2 and make two modifications, one a change of the independent variable from x to the current variable t, and the other a change due to the introduction of the second dependent variable in the present investigation.

To allow for the extra variable we separate the right-hand side of (9), § 8.2, into two parts. The first part, $F\,dx_b$, arises from the displacement of one end point B to B' along Γ_2, Fig. VIII. 1. The corresponding displacement here is from Q_m to Q_{m+1} along Γ_n and so $F\,dx_b$ must be replaced by $G\,dt$. The second part, $(g_2'-y')\dfrac{\partial F}{\partial y'}\,dx$, arises from the change in the value of the dependent variable y on passing from the extremal AB, Fig. VIII. 1, to the neighbouring curve AB'. The corresponding term for the present case is $(q-\dot{y})\dfrac{\partial G}{\partial \dot{y}}\,dt$, and comparison of Figs. VIII. 1 and

IX. 6 enables us to interpret q and $\dot{y}$ as follows. They are both rates of change of the ordinate of Q_m with respect to t, q as Q_m moves along Γ_n and $\dot{y}$ as Q_m moves along the extremal OQ_m.

The extra dependent variable is now easily allowed for by the addition of the term $(p-\dot{x})\dfrac{\partial G}{\partial \dot{x}}\,dt$, where the interpretations of p and $\dot{x}$ are as for q and y above with the word ordinate replaced by abscissa. If dt is the element of arc of Γ_n, then (p, q) are the direction cosines of the tangent to Γ_n at Q_m.

One point still requires investigation. The term

$$(g_2'-y')\frac{\partial F}{\partial y'}\,dx$$

in (9), § 8.2, was obtained with the help of the Eulerian equation (6), § 8.2. In the present case each of the variables x and y satisfies an Eulerian equation similar to (6), § 8.2, namely (2) and (3), § 9.9. The forms we have assumed, $(q-\dot{y})\dfrac{\partial G}{\partial \dot{y}}\,dt$ and $(p-\dot{x})\dfrac{\partial G}{\partial \dot{x}}\,dt$, are therefore justified.

Our final result is

$$I(OQ_{m+1})-I(OQ_m) = \left\{G+(p-\dot{x})\frac{\partial G}{\partial \dot{x}}+(q-\dot{y})\frac{\partial G}{\partial \dot{y}}\right\}dt, \qquad (2)$$

where terms of order $(dt)^2$ and smaller are neglected and the values of the variables are taken at the point Q_m.

Since $G(x, y, \dot{x}, \dot{y})$ is homogeneous and of degree one in $\dot{x}$ and $\dot{y}$, we may replace (2) by

$$I(OQ_{m+1})-I(OQ_m) = \left(p\frac{\partial G}{\partial \dot{x}}+q\frac{\partial G}{\partial \dot{y}}\right)dt. \qquad (3)$$

Evidently $\qquad I(Q_m Q_{m+1}) = G(x, y, p, q)\,dt, \qquad (4)$

where p and q have the same interpretations as for (2). Once again terms of order $(dt)^2$ and smaller have been neglected, as will be the case for the rest of this section.

Now write

$$E(x, y; \dot{x}, \dot{y}; p, q) = G(x, y, p, q)-p\frac{\partial G(x, y, \dot{x}, \dot{y})}{\partial \dot{x}}-q\frac{\partial G(x, y, \dot{x}, \dot{y})}{\partial \dot{y}}, \qquad (5)$$

then we have

$$I(OQ_m)+I(Q_m Q_{m+1})-I(OQ_{m+1}) = E(x,y;\dot{x},\dot{y};p,q)\,dt. \quad (6)$$

From (1) it follows immediately that

$$\delta I = \int\limits_A^B E(x,y;\dot{x},\dot{y};p,q)\,dt, \quad (7)$$

where the integration is along the curve Γ_n.

The function $E(x,y;\dot{x},\dot{y};p,q)$ is the Weierstrassian E function for the case when the integrand of I is expressed in parametric form. Since $G(x,y,\dot{x},\dot{y})$ is homogeneous and of degree one in $\dot{x}$, $\dot{y}$, it follows that $E(x,y;\dot{x},\dot{y};p,q)$ is homogeneous and of degree one in p, q and also homogeneous and of degree zero in $\dot{x},\dot{y}$. Thus, regarded as a function of $\dot{x}$ and $\dot{y}$, the E function must be of the form $\phi(\dot{y}/\dot{x})$ and so the individual values of $\dot{x}$ and $\dot{y}$ are immaterial as long as their ratio is unaltered. In E we may therefore take $(\dot{x},\dot{y})$ to be the direction cosines of the tangent to the extremal OQ_m at Q_m, Fig. IX. 6.

In the case where dt is the element of arc of Γ_n we may then replace $(\dot{x},\dot{y})$ by $(\cos\theta,\sin\theta)$ and (p,q) by $(\cos\phi, \sin\phi)$ in the E function, where $\tan\theta$ and $\tan\phi$ are the slopes at Q_m of OQ_m and Γ_n respectively. After these substitutions the E function usually assumes its simplest and most useful form.

9.11. Alternative forms for the E function

The E function can be expressed in a more convenient form by means of Taylor's theorem. If $f(x,y)$ has continuous partial derivatives of the second order, then†

$$f(x+h,y+k) = f(x,y)+\left(h\frac{\partial}{\partial x}+k\frac{\partial}{\partial y}\right)f(x,y)+$$

$$+\frac{1}{2}\left(h\frac{\partial}{\partial x_1}+k\frac{\partial}{\partial y_1}\right)^2 f(x_1,y_1) \quad (1)$$

where $x_1 = x+\lambda h$, $y_1 = y+\lambda k$, and $0<\lambda<1$. Applying this

† R. Courant, *Differential and Integral Calculus*, Blackie, vol. ii, p. 80.

theorem to $G(x,y,p,q)$, regarded as a function of p and q, we have

$$G(x,y,p,q) = G(x,y,\dot{x},\dot{y})+(p-\dot{x})\frac{\partial G}{\partial \dot{x}}+(q-\dot{y})\frac{\partial G}{\partial \dot{y}}+$$

$$+\frac{1}{2}\left\{(p-\dot{x})^2\frac{\partial^2}{\partial p_1^2}+2(p-\dot{x})(q-\dot{y})\frac{\partial^2}{\partial p_1\,\partial q_1}+\right.$$

$$\left.+(q-\dot{y})^2\frac{\partial^2}{\partial q_1^2}\right\}G(x,y,p_1,q_1), \qquad (2)$$

where $p_1 = \dot{x}+\lambda(p-\dot{x})$, $q_1 = \dot{y}+\lambda(q-\dot{y})$ and $0 < \lambda < 1$.

From the homogeneity property of $G(x,y,\dot{x},\dot{y})$ the first three terms on the right-hand side of (2) simplify to $p\dfrac{\partial G}{\partial \dot{x}}+q\dfrac{\partial G}{\partial \dot{y}}$.

On using (5), § 9.9, the remaining terms on the right of (2) simplify to

$$\tfrac{1}{2}(p\dot{y}-q\dot{x})^2S(x,y,p_1,q_1).$$

From the definition of E given by (5), § 9.10, it then follows that

$$E(x,y;\,\dot{x},\dot{y};\,p,q) = \tfrac{1}{2}(p\dot{y}-q\dot{x})^2S(x,y,p_1q_1), \qquad (3)$$

where p_1 and q_1 are as defined above.

When t is the length of arc of Γ_n we may write $\dot{x} = \cos\theta$, $p = \cos\phi$, where $\tan\theta$ and $\tan\phi$ are respectively the slopes of the extremal and of Γ_n. We now prove that

$$E(x,y;\,\cos\theta,\sin\theta;\,\cos\phi,\sin\phi)$$

$$= 2\sin^2\tfrac{1}{2}(\theta-\phi)S(x,y,\cos\bar{\theta},\sin\bar{\theta}), \qquad (4)$$

where $\bar{\theta}$ lies between† θ and ϕ.

From (5), § 9.10, and the homogeneity property of $G(x,y,\dot{x},\dot{y})$ we have

$$E(x,y;\,\cos\theta,\sin\theta;\,\cos\phi,\sin\phi)$$

$$= \cos\phi\left\{\frac{\partial G(x,y,p,q)}{\partial p}-\frac{\partial G(x,y,\dot{x},\dot{y})}{\partial \dot{x}}\right\}+$$

$$+\sin\phi\left\{\frac{\partial G(x,y,p,q)}{\partial q}-\frac{\partial G(x,y,\dot{x},\dot{y})}{\partial \dot{y}}\right\}, \qquad (5)$$

where, after the differentiations have been performed, we must write $\dot{x} = \cos\theta$, $\dot{y} = \sin\theta$, $p = \cos\phi$, and $q = \sin\phi$.

† (4) cannot be deduced from (3) by writing $\dot{x} = \cos\theta$, etc.

Now the coefficient of $\cos\phi$ is equal to

$$\int\limits_{\theta}^{\phi} \frac{d}{d\psi}\left\{\frac{\partial G(x,y,u,v)}{\partial u}\right\} d\psi, \tag{6}$$

where, after the partial differentiation with respect to u and before the differentiation with respect to ψ, we write $u = \cos\psi$ and $v = \sin\psi$. On differentiating with respect to ψ and using (5), § 9.9, the integrand of (6) can be expressed in terms of $S(x, y, \cos\psi, \sin\psi)$. We then prove that (6) is equal to

$$-\int\limits_{\theta}^{\phi} S(x,y,\cos\psi,\sin\psi)\sin\psi\, d\psi. \tag{7}$$

In the same way we can prove that the coefficient of $\sin\phi$ in (5) is equal to

$$\int\limits_{\theta}^{\phi} S(x,y,\cos\psi,\sin\psi)\cos\psi\, d\psi. \tag{8}$$

From (7) and (8) it follows that the right-hand side of (5) is equal to

$$\int\limits_{\theta}^{\phi} S(x,y,\cos\psi,\sin\psi)\sin(\phi-\psi)\, d\psi. \tag{9}$$

But by suitably measuring the angles θ and ϕ we can satisfy the inequality

$$-\pi < \theta-\phi \leqslant \pi$$

and so ensure that $\sin(\phi-\psi)$ does not change sign anywhere in the range of integration. We may then apply the mean-value theorem of the integral calculus† to (9) and deduce the relationship (4).

9.12. Conditions for maxima and minima of

$$I = \int\limits_{A}^{B} G(x,y,\dot{x},\dot{y})\, dt$$

THEOREM 19. *The following necessary and sufficient conditions ensure a strong minimum for I in the case where the end points A and B are fixed and where $G(x,y,\dot{x},\dot{y})$ is homogeneous and of degree one in $\dot{x}$ and $\dot{y}$.*

† R. Courant, *Differential and Integral Calculus*, vol. i, p. 128.

(i) *The variables x and y, which are functions of t, either satisfy the two equations*

$$\left.\begin{aligned} \frac{\partial G}{\partial x}-\frac{d}{dt}\left(\frac{\partial G}{\partial \dot{x}}\right) &= 0 \\ \frac{\partial G}{\partial y}-\frac{d}{dt}\left(\frac{\partial G}{\partial \dot{y}}\right) &= 0 \end{aligned}\right\} \tag{1}$$

or satisfy the single equivalent equation (9), § 9.9.

The curve whose equations in parametric form satisfy (1) *and which passes through the fixed end points A and B will be denoted by Γ_e.*

(ii) *The arc AB of Γ_e contains no points conjugate either to A or to B* (see § 2.6). *Singular points must also be excluded* (see § 9.9).

(iii) *At all points of this arc*

$$S(x, y, \dot{x}, \dot{y}) > 0, \tag{2}$$

where the function S is defined by (5), § 9.9.

(iv) *At all points of the arc*

$$E(x, y; \dot{x}, \dot{y}; p, q) > 0 \tag{3}$$

for every pair of finite values of p and q, other than $p = \dot{x}$ and $q = \dot{y}$ simultaneously.

If dt is the element of arc of Γ_n, see § 9.10, *then* (3) *can be replaced by*

$$E(x, y; \cos\theta, \sin\theta; \cos\phi, \sin\phi) > 0, \tag{3 a}$$

where $\tan\theta$ is the slope of the extremal. This inequality must be satisfied at every point of the arc AB and for all finite values of ϕ.

For a strong maximum the signs of the inequalities must be reversed.

To prove this we take Γ_e to be the extremal and Γ_n to be a neighbouring curve, as in § 9.9, but for simplicity we make $\epsilon_1 = \epsilon_2$. Since $\xi_1(t)$ and $\xi_2(t)$ are arbitrary functions of t, this entails no loss of generality.

From the analysis of § 9.9 and condition (i) it follows that the first variation, i.e. the term in δI containing ϵ_1 to the first degree, must vanish.

We continue for the moment with the case of weak variations. Here, if Q on Γ_n tends to P on Γ_e as ϵ tends to zero, then the slope of Γ_n at Q tends to that of Γ_e at P. Hence $p-\dot{x}$ and $q-\dot{y}$ must both tend to zero with ϵ. Thus, for sufficiently small ϵ we have

$$\left.\begin{aligned} p-\dot{x} &= k_1\epsilon+O(\epsilon^2) \\ q-\dot{y} &= k_2\epsilon+O(\epsilon^2) \end{aligned}\right\}, \tag{4}$$

where k_1 and k_2 are independent of ϵ. From (7), § 9.10, and (3), § 9.11, we deduce that

$$\delta I = \tfrac{1}{2}\epsilon^2 \int\limits_A^B (k_1\dot{y}-k_2\dot{x})^2 S(x,y,\dot{x},\dot{y})\, dt, \tag{5}$$

neglecting terms of order ϵ^3 and smaller. Hence condition (iii), in conjunction with (i), ensures a weak minimum for I. As in §§ 2.6 and 2.7, the permissible length of arc for such minimum is governed by the properties of conjugate points. Condition (ii) then allows a weak minimum for the whole length of arc AB.

Reverting to the case of strong variations, we observe that $S(x,y,\dot{x},\dot{y})$ plays in parametric theory a part analogous to that played by $\partial^2F/\partial y'^2$ in the analysis of § 2.4. If conditions (ii) and (iii) are fulfilled it can be proved, by arguments similar to those of § 9.5, that a field of extremals can be constructed of which Γ_e is a member. At every point of such a field $\dot{x}$ and $\dot{y}$ have unique values and therefore the integral of (7), § 9.10, can be evaluated for any given path Γ_n in the field. It follows immediately from (7), § 9.10, that condition (iv) is necessary and sufficient to ensure a strong minimum for I.

THEOREM 20. *The following are sufficient conditions to ensure a strong minimum for I in the case where the end points, A and B, are fixed and where $G(x,y,\dot{x},\dot{y})$ is homogeneous and of degree one in $\dot{x}$ and $\dot{y}$.*

(i) *and* (ii) *as in Theorem* 19.

(iii *a*) $$S(x,y,p,q) > 0 \tag{6}$$

at every point of AB and for every pair of values of p and q.

For a strong maximum the sign of the inequality (6) *must be reversed.*

To prove this it is evident that condition (iii), Theorem 19, is a special case of (iii *a*). Also, from (3), § 9.11, it follows that condition (iv), Theorem 19, is satisfied if (iii a) is satisfied. Therefore the conditions are sufficient to ensure the truth of Theorem 20.

But it is possible for conditions (iii) and (iv), Theorem 19, to be satisfied without (iii a) being true for *all* values of p and q. Thus the conditions of Theorem 20 although sufficient are not necessary.

9.13. Applications to special cases

Applying (5) § 9.9 and (5) § 9.10 to integrals of the type

$$I = \int_A^B g(x,y)(\dot{x}^2+\dot{y}^2)^{\frac{1}{2}}\, dt, \tag{1}$$

where the positive value of the root is taken, we have

$$S(x,y,\dot{x},\dot{y}) = \frac{g(x,y)}{(\dot{x}^2+\dot{y}^2)^{\frac{3}{2}}}, \tag{2}$$

$$E(x,y;\dot{x},\dot{y};p,q) = \frac{g(x,y)}{(\dot{x}^2+\dot{y}^2)^{\frac{1}{2}}}\{(p^2+q^2)^{\frac{1}{2}}(\dot{x}^2+\dot{y}^2)^{\frac{1}{2}}-p\dot{x}-q\dot{y}\}, \tag{3}$$

$$E(x,y;\cos\theta,\sin\theta;\cos\phi,\sin\phi) = \{1-\cos(\theta-\phi)\}g(x,y). \tag{4}$$

If $g(x,y) > 0$ the theorems of § 9.12 then show that I possesses a strong minimum when the path of integration is the arc of an extremal which contains no points conjugate to either end point.

Among the special cases of this result are (i) the shortest distance between two points, for which $I = \int (\dot{x}^2+\dot{y}^2)^{\frac{1}{2}}\, dt$ and $g(x,y) = 1$, (ii) the principle of least action for a particle for which $I = \int v(\dot{x}^2+\dot{y}^2)^{\frac{1}{2}}\, dt$, and (iii) the brachistochrone problem, $I = \int (1/v)(\dot{x}^2+\dot{y}^2)^{\frac{1}{2}}\, dt$, provided that v remains positive throughout the motion for (ii) and (iii).

9.14. Applications to geodesics on surfaces

Let (x,y,z) be the coordinates of a point P on a surface S and suppose that each coordinate can be expressed as a function of two variables u and v. If u and v are independent of each other their variations give us all points on S, but if they are functions of another variable t, then their variations give us points on curves lying wholly on S. We shall confine ourselves to a region which contains only one point corresponding to a given (u,v).

The distance from A to B, measured along a curve lying on S, is

$$I = \int_A^B (\dot{x}^2+\dot{y}^2+\dot{z}^2)^{\frac{1}{2}}\,dt \tag{1}$$

$$= \int_A^B (E\dot{u}^2+2F\dot{u}\dot{v}+G\dot{v}^2)^{\frac{1}{2}}\,dt, \tag{2}$$

where the positive value of the root is taken and where

$$\left.\begin{aligned} E &= \left(\frac{\partial x}{\partial u}\right)^2+\left(\frac{\partial y}{\partial u}\right)^2+\left(\frac{\partial z}{\partial u}\right)^2 \\ F &= \frac{\partial x}{\partial u}\frac{\partial x}{\partial v}+\frac{\partial y}{\partial u}\frac{\partial y}{\partial v}+\frac{\partial z}{\partial u}\frac{\partial z}{\partial v} \\ G &= \left(\frac{\partial x}{\partial v}\right)^2+\left(\frac{\partial y}{\partial v}\right)^2+\left(\frac{\partial z}{\partial v}\right)^2 \end{aligned}\right\}. \tag{3}$$

To avoid singular points on the arc AB we assume that $EG-F^2 > 0$. The geodesics of S are the paths of integration for which I is a minimum.

The Weierstrassian form of the Eulerian equation, (9), § 9.9, gives us, after some reduction,

$$(EG-F^2)(\dot{u}\ddot{v}-\dot{v}\ddot{u})+$$
$$+(E\dot{u}+F\dot{v})\left\{\left(\frac{\partial F}{\partial u}-\frac{1}{2}\frac{\partial E}{\partial v}\right)\dot{u}^2+\frac{\partial G}{\partial u}\dot{u}\dot{v}+\frac{1}{2}\frac{\partial G}{\partial v}\dot{v}^2\right\}-$$
$$-(F\dot{u}+G\dot{v})\left\{\frac{1}{2}\frac{\partial E}{\partial u}\dot{u}^2+\frac{\partial E}{\partial v}\dot{u}\dot{v}+\left(\frac{\partial F}{\partial v}-\frac{1}{2}\frac{\partial G}{\partial u}\right)\dot{v}^2\right\}=0. \tag{4}$$

In spite of its complexity this equation has a simple geometrical interpretation in terms of the concept known as geodesic curvature. For a plane curve let ψ be the angle between one of the axes and the tangent at a point P and let s be the length of arc measured from some convenient point to P. Then the rate $d\psi/ds$ measures the curvature at P. For the shortest distance between two points lying in the plane the curvature is zero at every point. This theory of curvature can be extended to curves lying on surfaces.

Let P be a point on a curve C lying on the surface S, then the geodesic through P which touches C is known as the geodesic tangent at P. Consider the curve on S defined by $u =$ constant. By varying this constant we obtain a family of curves which is, in many ways, analogous to the family of lines in a plane parallel to the x-axis. Through each point P passes only one member of this family. Let ψ denote the angle between this member and the geodesic tangent at P and s the length of arc of C measured from some convenient point to P. Then $d\psi/ds$ is the measure of the geodesic curvature of the curve C at P.

Equation (4) simply states that at every point of a geodesic the geodesic curvature is zero.†

The S function of (5), § 9.9, becomes for this case

$$S(u,v,\dot{u},\dot{v}) = \frac{EG-F^2}{(E\dot{u}^2+2F\dot{u}\dot{v}+G\dot{v}^2)^{\frac{3}{2}}}. \tag{5}$$

Since $EG-F^2 > 0$ and the positive value of the root is taken, the inequality (6), § 9.12, is satisfied. Hence, for geodesics the oonditions for a strong minimum are satisfied, subject, of course, to the requirements of conjugate points, (ii), § 9.12.

† A. R. Forsyth, *Lectures on the Differential Geometry of Curves and Surfaces*, Cambridge University Press, p. 153. The geodesic curvature in this book is denoted by $1/\gamma$, the parameters used are p and q (instead of u and v); and the symbols Δ and Γ are defined on p. 46.

INDEX

The numbers refer to chapters and sections

A CATALOG OF SELECTED

DOVER BOOKS

IN SCIENCE AND MATHEMATICS

A CATALOG OF SELECTED

DOVER BOOKS

IN SCIENCE AND MATHEMATICS

Astronomy

BURNHAM'S CELESTIAL HANDBOOK, Robert Burnham, Jr. Thorough guide to the stars beyond our solar system. Exhaustive treatment. Alphabetical by constellation: Andromeda to Cetus in Vol. 1; Chamaeleon to Orion in Vol. 2; and Pavo to Vulpecula in Vol. 3. Hundreds of illustrations. Index in Vol. 3. 2,000pp. 6⅛ x 9¼.
23567-X, 23568-8, 23673-0 Pa., Three-vol. set $46.85

THE EXTRATERRESTRIAL LIFE DEBATE, 1750–1900, Michael J. Crowe. First detailed, scholarly study in English of the many ideas that developed between 1750 and 1900 regarding the existence of intelligent extraterrestrial life. Examines ideas of Kant, Herschel, Voltaire, Percival Lowell, many other scientists and thinkers. 16 illustrations. 704pp. 5⅜ x 8½. 40675-X Pa. $19.95

A HISTORY OF ASTRONOMY, A. Pannekoek. Well-balanced, carefully reasoned study covers such topics as Ptolemaic theory, work of Copernicus, Kepler, Newton, Eddington's work on stars, much more. Illustrated. References. 521pp. 5⅜ x 8½.
65994-1 Pa. $15.95

AMATEUR ASTRONOMER'S HANDBOOK, J. B. Sidgwick. Timeless, comprehensive coverage of telescopes, mirrors, lenses, mountings, telescope drives, micrometers, spectroscopes, more. 189 illustrations. 576pp. 5⅜ x 8¼. (Available in U.S. only)
24034-7 Pa. $13.95

STARS AND RELATIVITY, Ya. B. Zel'dovich and I. D. Novikov. Vol. 1 of *Relativistic Astrophysics* by famed Russian scientists. General relativity, properties of matter under astrophysical conditions, stars and stellar systems. Deep physical insights, clear presentation. 1971 edition. References. 544pp. 5⅜ x 8½.
69424-0 Pa. $14.95

Chemistry

A SHORT HISTORY OF CHEMISTRY (3rd edition), J. R. Partington. Classic exposition explores origins of chemistry, alchemy, early medical chemistry, nature of atmosphere, theory of valency, laws and structure of atomic theory, much more. 428pp. 5⅜ x 8½. (Available in U.S. only) 65977-1 Pa. $12.95

CHEMICAL MAGIC, Leonard A. Ford. Second Edition, Revised by E. Winston Grundmeier. Over 100 unusual stunts demonstrating cold fire, dust explosions, much more. Text explains scientific principles and stresses safety precautions. 128pp. 5⅜ x 8½. 67628-5 Pa. $5.95

THE DEVELOPMENT OF MODERN CHEMISTRY, Aaron J. Ihde. Authoritative history of chemistry from ancient Greek theory to 20th-century innovation. Covers major chemists and their discoveries. 209 illustrations. 14 tables. Bibliographies. Indices. Appendices. 851pp. 5⅜ x 8½. 64235-6 Pa. $24.95

CATALYSIS IN CHEMISTRY AND ENZYMOLOGY, William P. Jencks. Exceptionally clear coverage of mechanisms for catalysis, forces in aqueous solution, carbonyl- and acyl-group reactions, practical kinetics, more. 864pp. 5⅜ x 8½. 65460-5 Pa. $19.95

THE HISTORICAL BACKGROUND OF CHEMISTRY, Henry M. Leicester. Evolution of ideas, not individual biography. Concentrates on formulation of a coherent set of chemical laws. 260pp. 5⅜ x 8½. 61053-5 Pa. $8.95

GENERAL CHEMISTRY, Linus Pauling. Revised 3rd edition of classic first-year text by Nobel laureate. Atomic and molecular structure, quantum mechanics, statistical mechanics, thermodynamics correlated with descriptive chemistry. Problems. 992pp. 5⅜ x 8½. 65622-5 Pa. $19.95

Engineering

DE RE METALLICA, Georgius Agricola. The famous Hoover translation of greatest treatise on technological chemistry, engineering, geology, mining of early modern times (1556). All 289 original woodcuts. 638pp. 6¾ x 11. 60006-8 Pa. $21.95

FUNDAMENTALS OF ASTRODYNAMICS, Roger Bate et al. Modern approach developed by U.S. Air Force Academy. Designed as a first course. Problems, exercises. Numerous illustrations. 455pp. 5⅜ x 8½. 60061-0 Pa. $12.95

DYNAMICS OF FLUIDS IN POROUS MEDIA, Jacob Bear. For advanced students of ground water hydrology, soil mechanics and physics, drainage and irrigation engineering and more. 335 illustrations. Exercises, with answers. 784pp. 6⅛ x 9¼. 65675-6 Pa. $19.95

ANALYTICAL MECHANICS OF GEARS, Earle Buckingham. Indispensable reference for modern gear manufacture covers conjugate gear-tooth action, gear-tooth profiles of various gears, many other topics. 263 figures. 102 tables. 546pp. 5⅜ x 8½. 65712-4 Pa. $16.95

ADVANCED STRENGTH OF MATERIALS, J. P. Den Hartog. Superbly written advanced text covers torsion, rotating disks, membrane stresses in shells, much more. Many problems and answers. 388pp. 5⅜ x 8½. 65407-9 Pa. $11.95

MECHANICS, J. P. Den Hartog. A classic introductory text or refresher. Hundreds of applications and design problems illuminate fundamentals of trusses, loaded beams and cables, etc. 334 answered problems. 462pp. 5⅜ x 8½. 60754-2 Pa. $12.95

MECHANICAL VIBRATIONS, J. P. Den Hartog. Classic textbook offers lucid explanations and illustrative models, applying theories of vibrations to a variety of practical industrial engineering problems. Numerous figures. 233 problems, solutions. Appendix. Index. Preface. 436pp. 5⅜ x 8½. 64785-4 Pa. $13.95

STRENGTH OF MATERIALS, J. P. Den Hartog. Full, clear treatment of basic material (tension, torsion, bending, etc.) plus advanced material on engineering methods, applications. 350 answered problems. 323pp. 5⅜ x 8½. 60755-0 Pa. $10.95

Mathematics

HANDBOOK OF MATHEMATICAL FUNCTIONS WITH FORMULAS, GRAPHS, AND MATHEMATICAL TABLES, edited by Milton Abramowitz and Irene A. Stegun. Vast compendium: 29 sets of tables, some to as high as 20 places. 1,046pp. 8 x 10½. 61272-4 Pa. $29.95

CALCULUS REFRESHER FOR TECHNICAL PEOPLE, A. Albert Klaf. Covers important aspects of integral and differential calculus via 756 questions. 566 problems, most answered. 431pp. 5⅜ x 8½. 20370-0 Pa. $9.95

ASYMPTOTIC EXPANSIONS OF INTEGRALS, Norman Bleistein & Richard A. Handelsman. Best introduction to important field with applications in a variety of scientific disciplines. New preface. Problems. Diagrams. Tables. Bibliography. Index. 448pp. 5⅜ x 8½. 65082-0 Pa. $13.95

FAMOUS PROBLEMS OF GEOMETRY AND HOW TO SOLVE THEM, Benjamin Bold. Squaring the circle, trisecting the angle, duplicating the cube: learn their history, why they are impossible to solve, then solve them yourself. 128pp. 5⅜ x 8½. 24297-8 Pa. $5.95

VECTOR AND TENSOR ANALYSIS WITH APPLICATIONS, A. I. Borisenko and I. E. Tarapov. Concise introduction. Worked-out problems, solutions, exercises. 257pp. 5⅝ x 8¼. 63833-2 Pa. $9.95

THE ABSOLUTE DIFFERENTIAL CALCULUS (CALCULUS OF TENSORS), Tullio Levi-Civita. Great 20th-century mathematician's classic work on material necessary for mathematical grasp of theory of relativity. 452pp. 5⅜ x 8½. 63401-9 Pa. $11.95

AN INTRODUCTION TO ORDINARY DIFFERENTIAL EQUATIONS, Earl A. Coddington. A thorough and systematic first course in elementary differential equations for undergraduates in mathematics and science, with many exercises and problems (with answers). Index. 304pp. 5⅜ x 8½. 65942-9 Pa. $9.95

FOURIER SERIES AND ORTHOGONAL FUNCTIONS, Harry F. Davis. An incisive text combining theory and practical example to introduce Fourier series, orthogonal functions and applications of the Fourier method to boundary-value problems. 570 exercises. Answers and notes. 416pp. 5⅜ x 8½. 65973-9 Pa. $13.95

COMPUTABILITY AND UNSOLVABILITY, Martin Davis. Classic graduate-level introduction to theory of computability, usually referred to as theory of recurrent functions. New preface and appendix. 288pp. 5⅜ x 8½. 61471-9 Pa. $8.95

ASYMPTOTIC METHODS IN ANALYSIS, N. G. de Bruijn. An inexpensive, comprehensive guide to asymptotic methods–the pioneering work that teaches by explaining worked examples in detail. Index. 224pp. 5⅜ x 8½. 64221-6 Pa. $7.95

ESSAYS ON THE THEORY OF NUMBERS, Richard Dedekind. Two classic essays by great German mathematician: on the theory of irrational numbers; and on transfinite numbers and properties of natural numbers. 115pp. 5⅜ x 8½. 21010-3 Pa. $6.95

THE GEOMETRY OF RENÉ DESCARTES, René Descartes. The great work founded analytical geometry. Original French text, Descartes's own diagrams, together with definitive Smith-Latham translation. 244pp. 5⅜ x 8½. 60068-8 Pa. $9.95

APPLIED COMPLEX VARIABLES, John W. Dettman. Step-by-step coverage of fundamentals of analytic function theory–plus lucid exposition of five important applications: Potential Theory; Ordinary Differential Equations; Fourier Transforms; Laplace Transforms; Asymptotic Expansions. 66 figures. Exercises at chapter ends. 512pp. 5⅜ x 8½. 64670-X Pa. $14.95

INTRODUCTION TO LINEAR ALGEBRA AND DIFFERENTIAL EQUATIONS, John W. Dettman. Excellent text covers complex numbers, determinants, orthonormal bases, Laplace transforms, much more. Exercises with solutions. Undergraduate level. 416pp. 5⅜ x 8½. 65191-6 Pa. $11.95

MATHEMATICAL METHODS IN PHYSICS AND ENGINEERING, John W. Dettman. Algebraically based approach to vectors, mapping, diffraction, other topics in applied math. Also generalized functions, analytic function theory, more. Exercises. 448pp. 5⅜ x 8¼. 65649-7 Pa. $12.95

THE THIRTEEN BOOKS OF EUCLID'S ELEMENTS, translated with introduction and commentary by Sir Thomas L. Heath. Definitive edition. Textual and linguistic notes, mathematical analysis. 2,500 years of critical commentary. Unabridged. 1,414pp. 5⅜ x 8½. Three-vol. set. Vol. I: 60088-2 Pa. $10.95
Vol. II: 60089-0 Pa. $10.95
Vol. III: 60090-4 Pa. $12.95

CALCULUS OF VARIATIONS WITH APPLICATIONS, George M. Ewing. Applications-oriented introduction to variational theory develops insight and promotes understanding of specialized books, research papers. Suitable for advanced undergraduate/graduate students as primary, supplementary text. 352pp. 5⅜ x 8½. 64856-7 Pa. $9.95

COMPLEX VARIABLES, Francis J. Flanigan. Unusual approach, delaying complex algebra till harmonic functions have been analyzed from real variable viewpoint. Includes problems with answers. 364pp. 5⅜ x 8½. 61388-7 Pa. $10.95

AN INTRODUCTION TO THE CALCULUS OF VARIATIONS, Charles Fox. Graduate-level text covers variations of an integral, isoperimetrical problems, least action, special relativity, approximations, more. References. 279pp. 5⅜ x 8½. 65499-0 Pa. $8.95

CATASTROPHE THEORY FOR SCIENTISTS AND ENGINEERS, Robert Gilmore. Advanced-level treatment describes mathematics of theory grounded in the work of Poincaré, R. Thom, other mathematicians. Also important applications to problems in mathematics, physics, chemistry and engineering. 1981 edition. References. 28 tables. 397 black-and-white illustrations. xvii + 666pp. 6⅛ x 9¼. 67539-4 Pa. $17.95

INTRODUCTION TO DIFFERENCE EQUATIONS, Samuel Goldberg. Exceptionally clear exposition of important discipline with applications to sociology, psychology, economics. Many illustrative examples; over 250 problems. 260pp. 5⅜ x 8½. 65084-7 Pa. $10.95

SPECIAL FUNCTIONS, N. N. Lebedev. Translated by Richard Silverman. Famous Russian work treating more important special functions, with applications to specific problems of physics and engineering. 38 figures. 308pp. 5⅜ x 8½. 60624-4 Pa. $9.95

QUALITATIVE THEORY OF DIFFERENTIAL EQUATIONS, V. V. Nemytskii and V.V. Stepanov. Classic graduate-level text by two prominent Soviet mathematicians covers classical differential equations as well as topological dynamics and ergodic theory. Bibliographies. 523pp. 5⅜ x 8½. 65954-2 Pa. $14.95

NUMBER THEORY AND ITS HISTORY, Oystein Ore. Unusually clear, accessible introduction covers counting, properties of numbers, prime numbers, much more. Bibliography. 380pp. 5⅜ x 8½. 65620-9 Pa. $10.95

THEORY OF MATRICES, Sam Perlis. Outstanding text covering rank, nonsingularity and inverses in connection with the development of canonical matrices under the relation of equivalence, and without the intervention of determinants. Includes exercises. 237pp. 5⅜ x 8½. 66810-X Pa. $8.95

OPTIMIZATION THEORY WITH APPLICATIONS, Donald A. Pierre. Broad spectrum approach to important topic. Classical theory of minima and maxima, calculus of variations, simplex technique and linear programming, more. Many problems, examples. 640pp. 5⅜ x 8½. 65205-X Pa. $17.95

INTRODUCTION TO ANALYSIS, Maxwell Rosenlicht. Unusually clear, accessible coverage of set theory, real number system, metric spaces, continuous functions, Riemann integration, multiple integrals, more. Wide range of problems. Undergraduate level. Bibliography. 254pp. 5⅜ x 8½. 65038-3 Pa. $9.95

MODERN NONLINEAR EQUATIONS, Thomas L. Saaty. Emphasizes practical solution of problems; covers seven types of equations. ". . . a welcome contribution to the existing literature...."–*Math Reviews*. 490pp. 5⅜ x 8½. 64232-1 Pa. $13.95

MATRICES AND LINEAR ALGEBRA, Hans Schneider and George Phillip Barker. Basic textbook covers theory of matrices and its applications to systems of linear equations and related topics such as determinants, eigenvalues and differential equations. Numerous exercises. 432pp. 5⅜ x 8½. 66014-1 Pa. $12.95

GEOMETRY OF COMPLEX NUMBERS, Hans Schwerdtfeger. Illuminating, widely praised book on analytic geometry of circles, the Moebius transformation, and two-dimensional non-Euclidean geometries. 200pp. 5⅝ x 8¼. 63830-8 Pa. $8.95

MATHEMATICS APPLIED TO CONTINUUM MECHANICS, Lee A. Segel. Analyzes models of fluid flow and solid deformation. For upper-level math, science and engineering students. 608pp. 5⅜ x 8½. 65369-2 Pa. $14.95

ELEMENTS OF REAL ANALYSIS, David A. Sprecher. Classic text covers fundamental concepts, real number system, point sets, functions of a real variable, Fourier series, much more. Over 500 exercises. 352pp. 5⅜ x 8½. 65385-4 Pa. $11.95

AN INTRODUCTION TO MATRICES, SETS AND GROUPS FOR SCIENCE STUDENTS, G. Stephenson. Concise, readable text introduces sets, groups, and most importantly, matrices to undergraduate students of physics, chemistry, and engineering. Problems. 164pp. 5⅜ x 8½. 65077-4 Pa. $7.95

SET THEORY AND LOGIC, Robert R. Stoll. Lucid introduction to unified theory of mathematical concepts. Set theory and logic seen as tools for conceptual understanding of real number system. 496pp. 5⅜ x 8¼. 63829-4 Pa. $14.95

LECTURES ON CLASSICAL DIFFERENTIAL GEOMETRY, Second Edition, Dirk J. Struik. Excellent brief introduction covers curves, theory of surfaces, fundamental equations, geometry on a surface, conformal mapping, other topics. Problems. 240pp. 5⅜ x 8½. 65609-8 Pa. $9.95

ORDINARY DIFFERENTIAL EQUATIONS, Morris Tenenbaum and Harry Pollard. Exhaustive survey of ordinary differential equations for undergraduates in mathematics, engineering, science. Thorough analysis of theorems. Diagrams. Bibliography. Index. 818pp. 5⅜ x 8½. 64940-7 Pa. $19.95

INTEGRAL EQUATIONS, F. G. Tricomi. Authoritative, well-written treatment of extremely useful mathematical tool with wide applications. Volterra Equations, Fredholm Equations, much more. Advanced undergraduate to graduate level. Exercises. Bibliography. 238pp. 5⅜ x 8½. 64828-1 Pa. $8.95

FOURIER SERIES, Georgi P. Tolstov. Translated by Richard A. Silverman. A valuable addition to the literature on the subject, moving clearly from subject to subject and theorem to theorem. 107 problems, answers. 336pp. 5⅜ x 8½. 63317-9 Pa. $11.95

DISTRIBUTION THEORY AND TRANSFORM ANALYSIS: An Introduction to Generalized Functions, with Applications, A. H. Zemanian. Provides basics of distribution theory, describes generalized Fourier and Laplace transformations. Numerous problems. 384pp. 5⅜ x 8½. 65479-6 Pa. $13.95

TENSOR CALCULUS, J.L. Synge and A. Schild. Widely used introductory text covers spaces and tensors, basic operations in Riemannian space, non-Riemannian spaces, etc. 324pp. 5⅜ x 8¼. 63612-7 Pa. $11.95

CALCULUS OF VARIATIONS, Robert Weinstock. Basic introduction covering isoperimetric problems, theory of elasticity, quantum mechanics, electrostatics, etc. Exercises throughout. 326pp. 5⅜ x 8½. 63069-2 Pa. $9.95

THE CONTINUUM: A Critical Examination of the Foundation of Analysis, Hermann Weyl. Classic of 20th-century foundational research deals with the conceptual problem posed by the continuum. 156pp. 5⅜ x 8½. 67982-9 Pa. $8.95

CHALLENGING MATHEMATICAL PROBLEMS WITH ELEMENTARY SOLUTIONS, A. M. Yaglom and I. M. Yaglom. Over 170 challenging problems on probability theory, combinatorial analysis, points and lines, topology, convex polygons, many other topics. Solutions. Total of 445pp. 5⅜ x 8½. Two-vol. set.
Vol. I: 65536-9 Pa. $8.95
Vol. II: 65537-7 Pa. $7.95

A SURVEY OF NUMERICAL MATHEMATICS, David M. Young and Robert Todd Gregory. Broad self-contained coverage of computer-oriented numerical algorithms for solving various types of mathematical problems in linear algebra, ordinary and partial, differential equations, much more. Exercises. Total of 1,248pp. 5⅜ x 8½. Two volumes.
Vol. I: 65691-8 Pa. $16.95
Vol. II: 65692-6 Pa. $16.95

INTRODUCTION TO PARTIAL DIFFERENTIAL EQUATIONS WITH APPLICATIONS, E. C. Zachmanoglou and Dale W. Thoe. Essentials of partial differential equations applied to common problems in engineering and the physical sciences. Problems and answers. 416pp. 5⅜ x 8½. 65251-3 Pa. $11.95

THE THEORY OF GROUPS, Hans J. Zassenhaus. Well-written graduate-level text acquaints reader with group-theoretic methods and demonstrates their usefulness in mathematics. Axioms, the calculus of complexes, homomorphic mapping, *p*-group theory, more. Many proofs shorter and more transparent than older ones. 276pp. 5⅜ x 8½. 40922-8 Pa. $12.95

GENERALIZED INTEGRAL TRANSFORMATIONS, A.H. Zemanian. Graduate-level study of recent generalizations of the Laplace, Mellin, Hankel, K. Weierstrass, convolution and other simple transformations. Bibliography. 320pp. 5⅜ x 8½. 65375-7 Pa. $8.95

Math–Decision Theory, Statistics, Probability

ELEMENTARY DECISION THEORY, Herman Chernoff and Lincoln E. Moses. Clear introduction to statistics and statistical theory covers data processing, probability and random variables, testing hypotheses, much more. Exercises. 364pp. 5⅜ x 8½. 65218-1 Pa. $12.95

STATISTICS MANUAL, Edwin L. Crow et al. Comprehensive, practical collection of classical and modern methods prepared by U.S. Naval Ordnance Test Station. Stress on use. Basics of statistics assumed. 288pp. 5⅜ x 8½. 60599-X Pa. $8.95

SOME THEORY OF SAMPLING, William Edwards Deming. Analysis of the problems, theory and design of sampling techniques for social scientists, industrial managers and others who find statistics increasingly important in their work. 61 tables. 90 figures. xvii + 602pp. 5⅜ x 8½. 64684-X Pa. $16.95

STATISTICAL ADJUSTMENT OF DATA, W. Edwards Deming. Introduction to basic concepts of statistics, curve fitting, least squares solution, conditions without parameter, conditions containing parameters. 26 exercises worked out. 271pp. 5⅜ x 8½. 64685-8 Pa. $9.95

LINEAR PROGRAMMING AND ECONOMIC ANALYSIS, Robert Dorfman, Paul A. Samuelson and Robert M. Solow. First comprehensive treatment of linear programming in standard economic analysis. Game theory, modern welfare economics, Leontief input-output, more. 525pp. 5⅜ x 8½. 65491-5 Pa. $17.95

DICTIONARY/OUTLINE OF BASIC STATISTICS, John E. Freund and Frank J. Williams. A clear concise dictionary of over 1,000 statistical terms and an outline of statistical formulas covering probability, nonparametric tests, much more. 208pp. 5⅜ x 8½. 66796-0 Pa.$8.95

PROBABILITY: An Introduction, Samuel Goldberg. Excellent basic text covers set theory, probability theory for finite sample spaces, binomial theorem, much more. 360 problems. Bibliographies. 322pp. 5⅜ x 8½. 65252-1 Pa. $10.95

GAMES AND DECISIONS: Introduction and Critical Survey, R. Duncan Luce and Howard Raiffa. Superb nontechnical introduction to game theory, primarily applied to social sciences. Utility theory, zero-sum games, n-person games, decision-making, much more. Bibliography. 509pp. 5⅜ x 8½. 65943-7 Pa. $14.95

FIFTY CHALLENGING PROBLEMS IN PROBABILITY WITH SOLUTIONS, Frederick Mosteller. Remarkable puzzlers, graded in difficulty, illustrate elementary and advanced aspects of probability. Detailed solutions. 88pp. 5⅜ x 8½. 65355-2 Pa. $4.95

PROBABILITY THEORY: A Concise Course, Y. A. Rozanov. Highly readable, self-contained introduction covers combination of events, dependent events, Bernoulli trials, etc. Translation by Richard Silverman. 148pp. 5⅜ x 8¼. 63544-9 Pa. $8.95

STATISTICAL METHOD FROM THE VIEWPOINT OF QUALITY CONTROL, Walter A. Shewhart. Important text explains regulation of variables, uses of statistical control to achieve quality control in industry, agriculture, other areas. 192pp. 5⅜ x 8½. 65232-7 Pa. $8.95

THE COMPLEAT STRATEGYST: Being a Primer on the Theory of Games of Strategy, J. D. Williams. Highly entertaining classic describes, with many illustrated examples, how to select best strategies in conflict situations. Prefaces. Appendices. 268pp. 5⅜ x 8½. 25101-2 Pa. $8.95

Math–History of

A SHORT ACCOUNT OF THE HISTORY OF MATHEMATICS, W. W. Rouse Ball. One of clearest, most authoritative surveys from the Egyptians and Phoenicians through 19th-century figures such as Grassman, Galois, Riemann. Fourth edition. 522pp. 5⅜ x 8½. 20630-0 Pa. $13.95

THE HISTORICAL ROOTS OF ELEMENTARY MATHEMATICS, Lucas N. H. Bunt, Phillip S. Jones, and Jack D. Bedient. Fundamental underpinnings of modern arithmetic, algebra, geometry and number systems derived from ancient civilizations. 320pp. 5⅜ x 8½. 25563-8 Pa. $9.95

GAMES, GODS & GAMBLING: A History of Probability and Statistical Ideas, F. N. David. Episodes from the lives of Galileo, Fermat, Pascal, and others illustrate this fascinating account of the roots of mathematics. Features thought-provoking references to classics, archaeology, biography, poetry. 1962 edition. 304pp. 5⅜ x 8½. (USO) 40023-9 Pa. $9.95

HISTORY OF MATHEMATICS, David E. Smith. Nontechnical survey from ancient Greece and Orient to late 19th century; evolution of arithmetic, geometry, trigonometry, calculating devices, algebra, the calculus. 362 illustrations. 1,355pp. 5⅜ x 8½. Two-vol. set.
Vol. I: 20429-4 Pa. $13.95
Vol. II: 20430-8 Pa. $14.95

A CONCISE HISTORY OF MATHEMATICS, Dirk J. Struik. The best brief history of mathematics. Stresses origins and covers every major figure from ancient Near East to 19th century. 41 illustrations. 195pp. 5⅜ x 8½. 60255-9 Pa. $8.95

Physics

OPTICAL RESONANCE AND TWO-LEVEL ATOMS, L. Allen and J. H. Eberly. Clear, comprehensive introduction to basic principles behind all quantum optical resonance phenomena. 53 illustrations. Preface. Index. 256pp. 5⅜ x 8½.
65533-4 Pa. $10.95

ULTRASONIC ABSORPTION: An Introduction to the Theory of Sound Absorption and Dispersion in Gases, Liquids and Solids, A. B. Bhatia. Standard reference in the field provides a clear, systematically organized introductory review of fundamental concepts for advanced graduate students, research workers. Numerous diagrams. Bibliography. 440pp. 5⅜ x 8½. 64917-2 Pa. $11.95

QUANTUM THEORY, David Bohm. This advanced undergraduate-level text presents the quantum theory in terms of qualitative and imaginative concepts, followed by specific applications worked out in mathematical detail. Preface. Index. 655pp. 5⅜ x 8½. 65969-0 Pa. $15.95

ATOMIC PHYSICS (8th edition), Max Born. Nobel laureate's lucid treatment of kinetic theory of gases, elementary particles, nuclear atom, wave-corpuscles, atomic structure and spectral lines, much more. Over 40 appendices, bibliography. 495pp. 5⅜ x 8½. 65984-4 Pa. $13.95

AN INTRODUCTION TO HAMILTONIAN OPTICS, H. A. Buchdahl. Detailed account of the Hamiltonian treatment of aberration theory in geometrical optics. Many classes of optical systems defined in terms of the symmetries they possess. Problems with detailed solutions. 1970 edition. xv + 360pp. 5⅜ x 8½.
67597-1 Pa. $10.95

HYDRODYNAMIC AND HYDROMAGNETIC STABILITY, S. Chandrasekhar. Lucid examination of the Rayleigh-Benard problem; clear coverage of the theory of instabilities causing convection. 704pp. 5⅜ x 8¼. 64071-X Pa. $17.95

INVESTIGATIONS ON THE THEORY OF THE BROWNIAN MOVEMENT, Albert Einstein. Five papers (1905–8) investigating dynamics of Brownian motion and evolving elementary theory. Notes by R. Fürth. 122pp. 5⅜ x 8½.
60304-0 Pa. $5.95

THE PHYSICS OF WAVES, William C. Elmore and Mark A. Heald. Unique overview of classical wave theory. Acoustics, optics, electromagnetic radiation, more. Ideal as classroom text or for self-study. Problems. 477pp. 5⅜ x 8½.
64926-1 Pa. $14.95

THIRTY YEARS THAT SHOOK PHYSICS: The Story of Quantum Theory, George Gamow. Lucid, accessible introduction to influential theory of energy and matter. Careful explanations of Dirac's anti-particles, Bohr's model of the atom, much more. 12 plates. Numerous drawings. 240pp. 5⅜ x 8½. 24895-X Pa. $7.95

ELECTRONIC STRUCTURE AND THE PROPERTIES OF SOLIDS: The Physics of the Chemical Bond, Walter A. Harrison. Innovative text offers basic understanding of the electronic structure of covalent and ionic solids, simple metals, transition metals and their compounds. Problems. 1980 edition. 582pp. 6⅛ x 9¼.
66021-4 Pa. $19.95